Springer Studies in Alternative Economics

This book series offers an outlet for cutting-edge research in alternative, heterodox and pluralist economics. It features scholarly studies on various schools of thought beyond the neo-classical orthodoxy, including Austrian, Post Keynesian, Sraffian, Marxian, Georgist, Institutional-evolutionary, as well as feminist, radical, social, green, and ecological economics. It aims to promote pluralism of economic ideas, methodological approaches, and topics.

The series also welcomes works that seek to develop alternative visions of the economy, economic structures and new approaches that aim to serve society, for example by embedding the economy within the ecosystem, or to enrich economic thought by advancing diversity, gender, race, and social equality. All titles in this series are peer-reviewed.

Jürgen Kremer

Capital Income and Concentration of Wealth

r > g and Redistribution

Springer

Jürgen Kremer
RheinAhrCampus
Koblenz University of Applied Sciences
Remagen, Rheinland-Pfalz, Germany

ISSN 2731-5908 ISSN 2731-5916 (electronic)
Springer Studies in Alternative Economics
ISBN 978-3-658-50714-5 ISBN 978-3-658-50715-2 (eBook)
https://doi.org/10.1007/978-3-658-50715-2

Previously published by Springer Fachmedien Wiesbaden

Modified and revised translation from the German language edition: "Kapitalerträge und Vermögenskonzentration" by Jürgen Kremer, © Der/die Herausgeber bzw. der/die Autor(en), exklusiv lizenziert an Springer Fachmedien Wiesbaden GmbH, ein Teil von Springer Nature 2024. Published by Springer Fachmedien Wiesbaden. All Rights Reserved.

This Springer imprint is published by the registered company Springer Fachmedien Wiesbaden GmbH, part of Springer Nature.
The registered company address is: Abraham-Lincoln-Str. 46, 65189 Wiesbaden, Germany

The flight of the aircraft over a run-down township is intended to illustrate the contrast between rich and poor

Preface

The content of this book pertains to the theory of economic growth, which by definition encompasses all theories that address the factors influencing a country's economic development—in both form and intensity. While there are various theories and approaches, they share the common feature of using gross domestic product (GDP) as an indicator of economic development. Another shared characteristic of growth theory is its long-term perspective, meaning that the focus lies on the development of economies over extended future periods.

For decades, growth theory has sought to explain growth endogenously, i.e., from within the respective model. This book marks a departure from this paradigm. Beginning with Chap. 2, we will not base our analysis on physical capital but will instead examine how the wealth of an economy as a whole and its household groups develop over time. Here, wealth will be interpreted in monetary terms and includes both financial and real assets. Instead of depreciation rates, as they appear in the standard growth models of the first chapter, capital return rates will be used and specified exogenously.

In addition, the conditions that influence the growth of an economy are considered so complex that no attempt will be made to describe them functionally. Research, innovation, technical progress, sufficient and affordable energy supplies, a functioning healthcare system, an efficient education system, a reliable legal framework, functioning state institutions, an intellectual atmosphere that promotes creativity and productivity, but also a stable peace order and functioning international trade in raw materials essential for production are complex factors that cannot be captured in simple formulas. Therefore, economic growth is also specified exogenously rather than endogenously in the following.

However, further analysis is then based on generally applicable balance equations that will be derived from circular flow models. The idea of circular flow models first appeared explicitly in the *Tableaux Économiques* of the physiocrats, which were developed by their most important representative, François Quesnay, in the eighteenth century.

The circular flow model for the overall economy will turn out to be a modified AK model in Sect. 3.2, and thus the circular flow models presented in this book are linked to one of the fundamental models of growth theory specified in [4].

The most important result of the extensive empirical research conducted by economist Thomas Piketty, which he published in his book *Capital in the 21st Century*, [17], is that he found a correlation between increasing economic inequality and capital concentration. This connection cannot be explained by standard growth models, but it does follow from an analysis of circular flow models.

To that end, the population is divided into a finite number of household groups, each with different average labor incomes, consumption expenditures, and wealth levels. Based on this structure, we examine the impact of key factors—namely, the economic growth rate and the rate of return on capital—not only on the long-term behavior of the economy as a whole but also on the development of the individual household groups within it. While positive economic growth rates, by definition, lead to GDP growth, it becomes evident that the economic development of individual household groups within the economy can vary significantly.

This book is divided into two parts. The first part examines continuous-time models. Chapter 1 introduces three classic growth models, the Solow-Swan model, the AK model, and the Schumpeterian model of creative destruction. Denoting the return on capital as r and the economic growth rate as g, then the relationship $r = g$ follows from both the AK and the Schumpeterian model and the relationship $r > g$ from the Solow-Swan model. When Piketty claimed a link between $r > g$ and increasing wealth concentration as the main result of his extensive empirical research, it sparked controversy and strong criticism. A major argument from critics was that the relation $r > g$ has long been known in economics, but no economist had ever claimed that this could explain growing inequality.

A central aim of this book is precisely to make that case. In Chap. 2, we will define the above mentioned circular flow models. Independently of the model, there are two sources of income: labor and capital income. While labor income compensates for one's own work, capital income is earned from owning capital. This means that these returns must be generated by other members of the economy. In Sect. 2.4, using an example, we show that capital income thus enables a redistribution mechanism, which we refer to in this book as interest transfer.

When the parameters of the circular flow model are assumed to be constant over time, as shown in Chap. 3, it is possible to derive closed-form expressions for all key economic variables, including wealth, labor income, and interest transfers for each household group. Based on these results, it is demonstrated that in a circular flow model with constant parameters, wealth concentration occurs for large times if $r \geq g \geq 0$ and $r > 0$, given that wealthier and less wealthy household groups exist from the beginning, as discussed in Sect. 3.4.

To quantify economic inequality, Chap. 4 introduces the concepts of Lorenz curves and the Gini coefficient, which are analyzed for the circular flow model with constant parameters.

The simple differential equations in Sect. 3.3, which describe the development of wealth for household groups, result from assuming time-invariant parameters for the corresponding circular flow model. If these parameters are allowed to change over time, the solution methods presented in Sect. 3.3 no longer apply.

However, time-varying model parameters are realistic, particularly if household groups are allowed to react to events. For instance, households may cut back on planned consumption to a certain extent before going into debt if they can no longer finance their expenses, or wealthier households may be favored in the distribution of total labor income if total labor income is insufficient to increase the labor income of each group equally, for example, in line with economic growth.

These examples show that it would be reasonable to react to events as they occur. To account for these, the circular flow models in the second part of the book, starting in Chap. 5, are reformulated from continuous time to discrete time. This allows the development of the household groups to be calculated iteratively, i.e., year by year, and at each point in time, it becomes possible to respond to events using rules.

In Chap. 6, we first demonstrate that the discrete-time circular flow model also has analytical solutions if the model parameters are assumed to be constant over time. The resulting formulas resemble those derived in Chap. 3 and lead to the same conclusions regarding wealth concentration.

Rules for the distribution of labor income among household groups, as well as for the development and allocation of consumption expenditures among these groups, are presented in Chap. 7.

Since the analyses in this book identify capital income as particularly decisive for wealth concentration, the question arises as to what extent a tax system could influence rising inequality. To explore this, Chap. 8 extends the discrete-time circular flow model to include the state and its tax system. In particular, flexible rules for taxing labor and capital income are modeled, and their effects are examined using simulations.

It is shown that heavy taxation of high capital returns can slow down the concentration of wealth, and that a reduction of the Gini coefficient for wealth may be achieved. However, this does not reverse the overall distribution of wealth within economies. Wealthier household groups remain wealthy over time, but the disparities in wealth are reduced.

I would like to thank Dr. Markus Grabka from the German Institute for Economic Research (DIW Berlin) for providing data on wealth and income distribution, and for allowing the reproduction of Fig. 7.1. I would also like to thank Dr. Karl Herweg for granting permission to use his two Monopoly cartoons in Fig. 8.11.

Daun, Germany
November 25, 2025

Jürgen Kremer

Competing Interests The author has no competing interests to declare that are relevant to the content of this manuscript.

Contents

Part I
Continuous-Time Circular Flow Models

Chapter 1
Introduction: Standard Growth Models

In their work [4] on growth theory, the two Nobel Prize-winning economists Philippe Aghion and Peter Howitt list four growth paradigms that they consider fundamental to growth theory:

- The neoclassical growth model
- The AK model
- The model with product diversity
- The Schumpeterian model

The neoclassical growth model, the AK model, and the Schumpeterian model, which can be interpreted as a further development of the model with product diversity, are outlined in this introductory chapter in Sects. 1.1, 1.2, and 1.3, respectively.

If we denote the rate of return on capital by r and economic growth by g, then the neoclassical growth model implies the relationship $r > g$. The AK model and the Schumpeterian model imply the relationship $r = g$. This relationship $r > g$ has long been known in economics and is not associated with increasing economic inequality, as explained in Sect. 1.4.

For more on the topics covered in this chapter, see also [5] by Barro and Sala-i-Martin.

1.1 The Neoclassical Growth Model

The neoclassical growth model was developed in 1956 by Robert Merton Solow and Trevor Swan and is therefore also known as the **Solow-Swan model**. This model uses a **production function** that represents the **output** Y of an economy as a function of the **labor** L, **capital** K and **technology** T employed

J. Kremer, *Capital Income and Concentration of Wealth*, Springer Studies in Alternative Economics, https://doi.org/10.1007/978-3-658-50715-2_1

$$Y = F(K, L, T). \tag{1.1}$$

In addition to certain requirements on the first and second partial derivatives of F with respect to K and L, production functions are assumed to be positively homogeneous of degree 1 in K and L, so that

$$F(aK, aL, T) = aF(K, L, T) \quad (a > 0). \tag{1.2}$$

Equation (1.2) means that the use of a-fold capital and labor leads to a-fold output. A production function frequently used in growth theory is the **Cobb-Douglas function**

$$F(K, L) = AK^{\alpha}L^{1-\alpha} \quad (0 < A, 0 < \alpha < 1). \tag{1.3}$$

If α is small, labor has a stronger influence on output. As α approaches one, capital becomes increasingly dominant.

We consider a closed economy whose industry only manufactures products from a single sector that can be both consumed and invested. Furthermore, a constant savings rate is assumed

$$s = \frac{S}{Y}, \tag{1.4}$$

as well as a constant **growth rate** $\lambda \geq 0$ **of the population**

$$\dot{L} = \lambda L,$$

so that

$$L(t) = L_0 e^{\lambda t}. \tag{1.5}$$

A dot above a variable denotes, here and in the following, the derivative of that variable with respect to time. Using a constant **depreciation rate** $\delta > 0$, the change in the capital stock over time is written as

$$\begin{aligned} \dot{K} &= S - \delta K \qquad (1.6) \\ &= sY - \delta K \\ &= sAK^{\alpha}L^{1-\alpha} - \delta K, \end{aligned}$$

where a Cobb-Douglas function was used in the last line of (1.6) to represent Y. This is a nonlinear differential equation for the physical capital stock. But if we employ the transformation $X = K^{1-\alpha}$, the differential equation (1.6) for K can be converted into a differential equation for X

$$\begin{aligned}\dot{X} &= (1-\alpha)\,K^{-\alpha}\dot{K} \\ &= (1-\alpha)\,K^{-\alpha}\left(sAK^{\alpha}L^{1-\alpha} - \delta K\right) \\ &= (1-\alpha)\,sAL^{1-\alpha} - (1-\alpha)\,\delta X.\end{aligned} \tag{1.7}$$

With

$$\begin{aligned}a &= (1-\alpha)\,\delta, \\ b &= (1-\alpha)\,sAL_0^{1-\alpha}, \\ c &= (1-\alpha)\,\lambda\end{aligned}$$

and (1.5), we may write (1.7) as

$$\dot{X} + aX = be^{ct}. \tag{1.8}$$

This is an ordinary linear inhomogeneous first-order differential equation for X. To find a special solution of (1.8), we plug in the ansatz function

$$X_1(t) = de^{ct}$$

and compute

$$cde^{ct} + ade^{ct} = be^{ct}$$

so that

$$d = \frac{b}{a+c} = \frac{sAL_0^{1-\alpha}}{\delta + \lambda}.$$

The general solution X_2 of the homogeneous differential equation belonging to (1.8),

$$\dot{X} + aX = 0,$$

is

$$X_2(t) = x_0 e^{-at},$$

where x_0 denotes an arbitrary constant. The general solution X of (1.8) is given by the sum of a special solution X_1 and the general solution X_2 of the corresponding homogeneous equation, so that

$$X(t) = X_1(t) + X_2(t) = de^{ct} + x_0 e^{-at}.$$

The inverse transformation $K = X^{\frac{1}{1-\alpha}}$ finally leads to a solution of (1.6)

$$K(t) = \left(de^{ct} + x_0 e^{-at}\right)^{\frac{1}{1-\alpha}} \tag{1.9}$$
$$= \left(\frac{sAL_0^{1-\alpha}}{\delta+\lambda} e^{(1-\alpha)\lambda t} + x_0 e^{-(1-\alpha)\delta t}\right)^{\frac{1}{1-\alpha}}$$
$$= \left(\frac{sAL_0^{1-\alpha}}{\delta+\lambda} e^{(1-\alpha)\lambda t} + \left(K_0^{1-\alpha} - \frac{sAL_0^{1-\alpha}}{\delta+\lambda}\right) e^{-(1-\alpha)\delta t}\right)^{\frac{1}{1-\alpha}}.$$

In the last line of (1.9),

$$x_0 = K_0^{1-\alpha} - \frac{sAL_0^{1-\alpha}}{\delta+\lambda}$$

was defined, so that we have $K(0) = K_0$. Further, the Cobb-Douglas function (1.3) together with the expressions (1.5) for L and (1.9) for K leads to a closed formula for the output Y.

To examine the per capita capital $k = K/L$, we first calculate

$$K(t) = \left(\frac{sA}{\delta+\lambda} L(t)^{1-\alpha} + \left(K_0^{1-\alpha} L_0^{-(1-\alpha)} - \frac{sA}{\delta+\lambda}\right) L(t)^{1-\alpha} e^{-(1-\alpha)(\delta+\lambda)t}\right)^{\frac{1}{1-\alpha}}$$
$$= L(t) \left(\frac{sA}{\delta+\lambda} + \left(\left(\frac{K_0}{L_0}\right)^{1-\alpha} - \frac{sA}{\delta+\lambda}\right) e^{-(1-\alpha)(\delta+\lambda)t}\right)^{\frac{1}{1-\alpha}}$$

and obtain with $k_0 = K_0/L_0$

$$k(t) = \frac{K(t)}{L(t)} = \left(\frac{sA}{\delta+\lambda} + \left(k_0^{1-\alpha} - \frac{sA}{\delta+\lambda}\right) e^{-(1-\alpha)(\delta+\lambda)t}\right)^{\frac{1}{1-\alpha}}. \tag{1.10}$$

Furthermore, with (1.3) we find

$$y = \frac{Y}{L} = Ak^{\alpha}, \tag{1.11}$$

and with (1.10) we may write the per capita output $y = Y/L$ as

$$y(t) = \left(\frac{sA^{\frac{1}{\alpha}}}{\delta+\lambda} + \left(y_0^{\frac{1-\alpha}{\alpha}} - \frac{sA^{\frac{1}{\alpha}}}{\delta+\lambda}\right) e^{-(1-\alpha)(\delta+\lambda)t}\right)^{\frac{\alpha}{1-\alpha}}, \tag{1.12}$$

where $y_0 = y(0) = Ak_0^{\alpha}$ has been used.

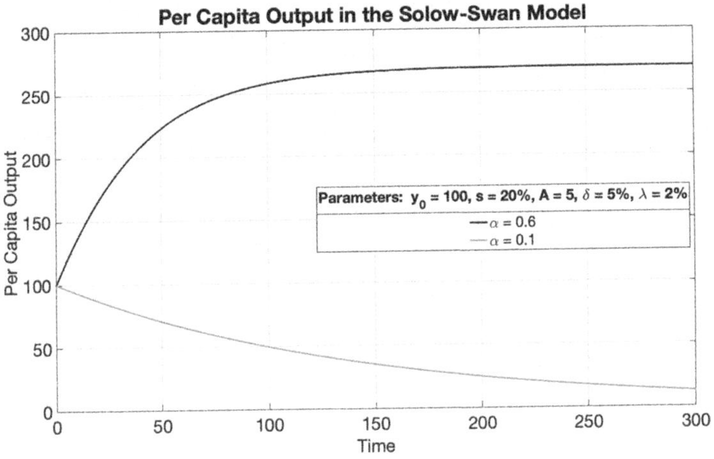

Fig. 1.1 Per capita output y in the Solow-Swan model for two parameter sets

From (1.10) and (1.12), we obtain the limits

$$\lim_{t\to\infty} k(t) = \left(\frac{sA}{\delta+\lambda}\right)^{\frac{1}{1-\alpha}}$$

and

$$\lim_{t\to\infty} y(t) = \left(\frac{sA^{\frac{1}{\alpha}}}{\delta+\lambda}\right)^{\frac{\alpha}{1-\alpha}}.$$

Thus, per capita capital $k(t)$ and output per capita $y(t)$ both converge to finite limits for $t \to \infty$. Therefore, eternal growth per capita is not possible in the basic Solow-Swan model. In Fig. 1.1, the time evolution of per capita output y is illustrated for two parameter sets.

Economic Growth and Capital Income

In the framework of the Solow-Swan model we now compare the growth rates for the economy and for capital income. With (1.10) and the abbreviations $a = \frac{sA}{\delta+\lambda} > 0$, $b = k_0^{1-\alpha} - \frac{sA}{\delta+\lambda}$, and $c = (1-\alpha)(\delta+\lambda) > 0$, we obtain $\lim_{t\to\infty} ae^{ct} = +\infty$ so that

$$\frac{\mathrm{d}}{\mathrm{d}t}\ln k = \frac{\mathrm{d}}{\mathrm{d}t}\ln\left(a+be^{-ct}\right)^{\frac{1}{1-\alpha}} \tag{1.13}$$

$$
\begin{aligned}
&= \frac{1}{1-\alpha}\frac{\mathrm{d}}{\mathrm{d}t}\ln\left(a + be^{-ct}\right) \\
&= \frac{1}{1-\alpha}\frac{-bce^{-ct}}{a+be^{-ct}} \\
&= -b\frac{c}{1-\alpha}\frac{1}{ae^{ct}+b} \\
&\to 0
\end{aligned}
$$

for $t \to \infty$. In addition, the sign of $\frac{\mathrm{d}}{\mathrm{d}t}\ln k$ matches the sign of $-b$ for t large enough.

Assuming

$$\frac{\dot{k}}{k} = \frac{\mathrm{d}}{\mathrm{d}t}\ln k > 0$$

and using (1.11), we conclude

$$\frac{\dot{y}}{y} = \frac{\mathrm{d}}{\mathrm{d}t}\ln y = \frac{\mathrm{d}}{\mathrm{d}t}\ln\left(Ak^{\alpha}\right) = \frac{\mathrm{d}}{\mathrm{d}t}\ln A + \alpha\frac{\mathrm{d}}{\mathrm{d}t}\ln k = \alpha\frac{\dot{k}}{k} < \frac{\dot{k}}{k},$$

so that

$$\frac{\mathrm{d}}{\mathrm{d}t}\ln y < \frac{\mathrm{d}}{\mathrm{d}t}\ln k. \tag{1.14}$$

We define **economic growth** g_t and the **return on capital** r_t as the **continuous returns** of the respective variables Y and K at time t. Thus, we obtain

$$g_t = \frac{\mathrm{d}}{\mathrm{d}t}\ln Y = \frac{\mathrm{d}}{\mathrm{d}t}\ln(yL) = \frac{\mathrm{d}}{\mathrm{d}t}(\ln y + \ln L) = \frac{\mathrm{d}}{\mathrm{d}t}\ln y + \lambda \tag{1.15}$$

and

$$r_t = \frac{\mathrm{d}}{\mathrm{d}t}\ln K = \frac{\mathrm{d}}{\mathrm{d}t}\ln(kL) = \frac{\mathrm{d}}{\mathrm{d}t}(\ln k + \ln L) = \frac{\mathrm{d}}{\mathrm{d}t}\ln k + \lambda. \tag{1.16}$$

With (1.14) we conclude

$$r_t > g_t. \tag{1.17}$$

The rate of return on capital r_t is therefore greater than economic growth g_t. With (1.13) we further have

$$\lim_{t\to\infty}\frac{\dot{y}}{y} = 0 = \lim_{t\to\infty}\frac{\dot{k}}{k}$$

and thus

$$\lim_{t \to \infty} g_t = \lim_{t \to \infty} r_t = \lambda.$$

If the continuous rate of return on per capita capital is positive, $\dot{k}/k > 0$, then the basic form of the Solow-Swan model implies that the rate of return on capital is greater than economic growth, even if both variables converge toward a common constant, the population growth rate, for long periods of time.

Monetization of the Solow-Swan Model

Barro and Sala-i-Martin show in [5] that eternal per capita growth can be realized in the Solow-Swan model if increasing levels of technology are assumed.

However, a monetary interpretation of the Solow-Swan model also leads to perpetual growth. If we examine the solutions for k and y in the original model (see Eqs. (1.10) and (1.12)), we observe that the depreciation factor δ appears in a critical way—and with the "wrong" sign.

Now we assume that capital K represents the economy's monetary wealth, rather than the quantity of goods produced. Since the money supply can theoretically be expanded indefinitely, there is no physical constraint on unlimited growth under this interpretation.

The change in total wealth $\dot{K}$ is now written analogously to Eq. (1.6) as the sum of savings S and the interest ρK on existing capital K

$$\dot{K} = S + \rho K,$$

where ρ denotes the interest rate. If all other assumptions of the Solow-Swan model remain unchanged, the solutions for capital and output per capita become

$$k(t) = \left(-\frac{sA}{\rho - \lambda} + \left(k_0^{1-\alpha} + \frac{sA}{\rho - \lambda}\right) e^{(1-\alpha)(\rho-\lambda)t}\right)^{\frac{1}{1-\alpha}}$$

and

$$y(t) = Ak^{\alpha}(t) = A\left(-\frac{sA}{\rho - \lambda} + \left(k_0^{1-\alpha} + \frac{sA}{\rho - \lambda}\right) e^{(1-\alpha)(\rho-\lambda)t}\right)^{\frac{\alpha}{1-\alpha}}$$

for $\rho \neq \lambda$. Thus, if the interest rate ρ exceeds the population growth rate λ, eternal per capita growth of capital occurs, see Fig. 1.2.

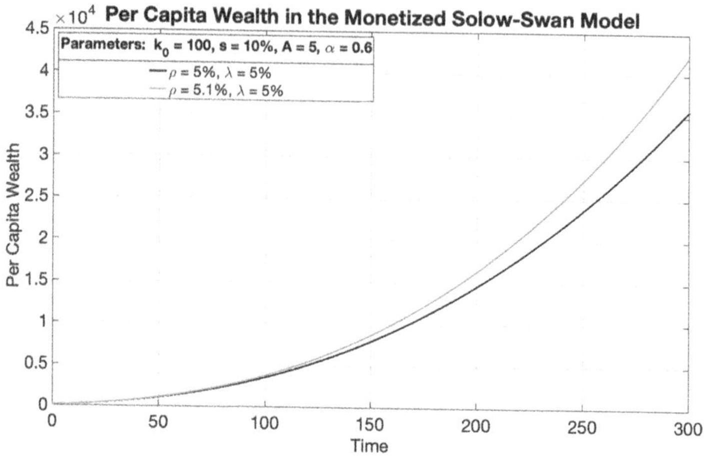

Fig. 1.2 Per capita wealth k in the monetized Solow-Swan model for two parameter sets

1.2 The AK Model

The AK model was developed by Marvin Frankel in 1962 and is considered the basis of endogenous growth theory. In the derivation, see [4], the production function

$$Y = F(K) = \left(AK^{\eta}\right) K^{\alpha} = AK^{\alpha+\eta} \quad (0 < \alpha + \eta)$$

is used. The factor K^{α} originally stems from an assumed Cobb-Douglas function, in which the number L of members of the economy was assumed to be constant. The factor K^{η} is intended to explain increasing productivity due to technologies that improve over time. Assuming again a constant savings rate $s = S/Y$ and a constant depreciation rate δ, the change in the capital stock over time is given by

$$\begin{aligned} \dot{K} &= S - \delta K \\ &= sY - \delta K \\ &= sAK^{\alpha+\eta} - \delta K. \end{aligned} \tag{1.18}$$

Thus, the growth rate $\dot{K}/K$ of the capital stock turns out to be

$$\frac{\dot{K}}{K} = sAK^{\alpha+\eta-1} - \delta.$$

We can now distinguish between the cases $\alpha + \eta < 1$, $\alpha + \eta > 1$, and $\alpha + \eta = 1$. In the first case, $\alpha + \eta < 1$, technological progress is not strong enough to counteract the declining returns of individual capital accumulation, and the growth rate $\dot{K}/K$

converges toward zero over time. In the case $\alpha + \eta > 1$, technological progress is so strong that the entire economy exhibits permanently rising growth rates, which is referred to in [4] as explosive growth and is considered unrealistic.

In the remaining case, $\alpha + \eta = 1$, the declining returns resulting from the Cobb-Douglas function are precisely offset by technological progress. Technological progress causes long-term, but not explosive, growth, and we obtain

$$Y = AK = F(K). \tag{1.19}$$

The model owes its name to this proportional relationship between output Y and capital K, given by (1.19).

Alternatively, the exponent $\alpha = 1$ could be chosen in the Cobb-Douglas function of the Solow-Swan model. Then the parameter L for labor can be suppressed, as it no longer appears in the production function in this case, and the Solow-Swan model immediately specializes to the AK model.

In any case, (1.18) reduces to

$$\dot{K} = (sA - \delta) K, \tag{1.20}$$

and with $K_0 = K(0)$ we conclude

$$K(t) = K_0 e^{(sA-\delta)t}$$

and

$$Y(t) = AK(t) = AK_0 e^{(sA-\delta)t}.$$

For economic growth g and for the capital return rate r we thus obtain

$$g = \frac{\mathrm{d}}{\mathrm{d}t} \ln Y = \frac{\dot{Y}}{Y} = sA - \delta, \tag{1.21}$$

$$r = \frac{\mathrm{d}}{\mathrm{d}t} \ln K = \frac{\dot{K}}{K} = sA - \delta.$$

In the AK model, economic growth and the rate of return on capital are therefore identical and have the constant value

$$g = r = sA - \delta. \tag{1.22}$$

Economic growth g and the rate of return on capital r increase with rising savings rates s, see Fig. 1.3.

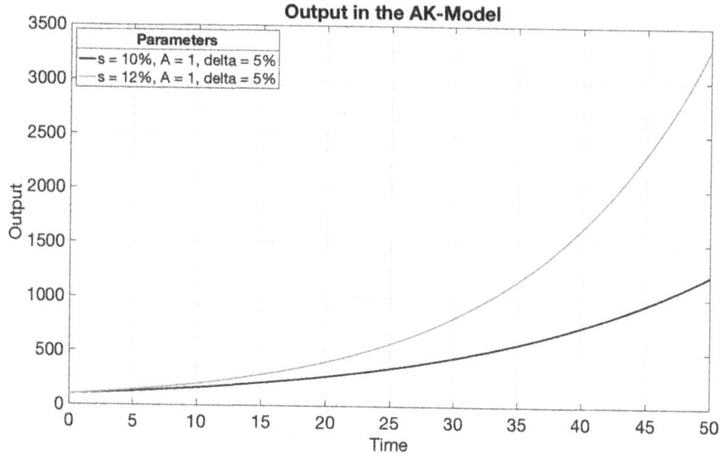

Fig. 1.3 Output Y in the AK model for two parameter sets

1.3 The Schumpeterian Model

The Schumpeterian model was formulated in 1992 by Philippe Aghion and Peter Howitt and is a further development of the product diversity model. While in the product diversity model, obsolete technologies continue to exist alongside new developments, in the Schumpeterian model, new, improved technologies replace the old ones. This model thus formalizes Joseph Schumpeter's concept of *creative destruction* and has the characteristics described in [4], Section 4.2:

There is a sequence of discrete time periods $t = 1, 2, \ldots$, and in each period there is a fixed number L of individuals, each offering one unit of labor. The members of the economy consume only one good, the so-called final product, which is produced using the labor L and a single intermediate product x_t according to the Cobb-Douglas production function

$$Y_t = (A_t L)^{1-\alpha} x_t^{\alpha}. \tag{1.23}$$

The variable Y_t is called the **final output**. Here, A_t denotes the **productivity of the intermediate product** in period t. The intermediate product is produced using the final product as the sole input in a 1:1 ratio. This means that for each unit of the intermediate product, the monopolist must use one unit of the final product as input. The final production that is not used for intermediate production is available for consumption and research and forms the **gross domestic product** G_t of the economy, so that

$$G_t = Y_t - x_t. \tag{1.24}$$

Production and Profits

The monopolist maximizes its expected consumption in period t by maximizing its **profit**

$$\Pi_t = p_t x_t - x_t, \tag{1.25}$$

measured in units of the final product. Here, p_t denotes the price of the intermediate product relative to the final product. The monopolist's revenue corresponds to $p_t x_t$, and its costs consist of the use of x_t in the final product. According to economic folklore, the equilibrium price of a production factor should correspond to the value of its marginal product, i.e., the following should apply:

$$p_t = \partial Y_t / \partial x_t = \alpha \left(A_t L\right)^{1-\alpha} x_t^{\alpha-1}.$$

Insertion of p_t into (1.25) leads to

$$\Pi_t = \alpha \left(A_t L\right)^{1-\alpha} x_t^{\alpha} - x_t.$$

Using the first-order condition $\partial \Pi_t / \partial x_t = 0$, we find

$$x_t = \alpha^{\frac{2}{1-\alpha}} A_t L, \tag{1.26}$$

leading to the optimal profit

$$\Pi_t = (1-\alpha)\, \alpha^{\frac{1+\alpha}{1-\alpha}} A_t L.$$

If (1.26) is inserted into (1.23) and in (1.24), we find the expressions

$$\begin{aligned} Y_t &= \alpha^{\frac{2\alpha}{1-\alpha}} A_t L \\ G_t &= \alpha^{\frac{2\alpha}{1-\alpha}} \left(1-\alpha^2\right) A_t L \end{aligned} \tag{1.27}$$

for final output and gross domestic product. Thus, we conclude

$$\begin{aligned} \Pi_t &= c_\Pi A_t, & c_\Pi &= (1-\alpha)\, \alpha^{\frac{1+\alpha}{1-\alpha}} L, \\ x_t &= c_X A_t, & c_X &= \alpha^{\frac{2}{1-\alpha}} L, \\ Y_t &= c_Y A_t, & c_Y &= \alpha^{\frac{2\alpha}{1-\alpha}} L, \\ G_t &= c_G A_t, & c_G &= \alpha^{\frac{2\alpha}{1-\alpha}} \left(1-\alpha^2\right) L, \end{aligned} \tag{1.28}$$

and all listed variables prove to be proportional to productivity A_t under the specified model assumptions.

Innovations

In each period, there is a person called an entrepreneur who has the opportunity to try out an innovation. If successful, the innovation creates a new version of the intermediate product that is more productive than previous versions. In this case, the modeling is done in such a way that the productivity of the intermediate product used in period t increases from the value A_{t-1} to the value $A_t = \gamma A_{t-1}$, where $\gamma > 1$ is assumed. However, if the innovation is unsuccessful, the intermediate product at time t is the same as at time $t-1$, i.e., $A_t = A_{t-1}$.

Research involves costs and research results are uncertain. It is assumed that the probability μ_t of a successful innovation increases with research expenditure R_t, which, however, is related to the respective successful productivity levels $A_t^* = \gamma A_{t-1}$. For this purpose, the ansatz

$$\mu_t = \phi\left(\frac{R_t}{A_t^*}\right)$$

is chosen with a probability function ϕ that is not initially specified in detail. In [4], Aghion and Howitt define $n_t = R_t/A_t^*$ and choose a Cobb-Douglas approach for ϕ

$$\phi(n) = \lambda n^{\sigma}, \tag{1.29}$$

where they interpret λ as a parameter that reflects the **productivity of the research sector** and they denote $0 < \sigma < 1$ as production elasticity. Using this approach, we obtain $\phi'(n) > 0$ and $\phi''(n) < 0$.

Research Expenditures

If an entrepreneur can successfully implement an innovation in period t, then he can incur the profit

$$\Pi_t^* = c_\Pi A_t^*. \tag{1.30}$$

The reward for a successful innovation is therefore Π_t^*, and since the innovation has a probability of success of μ_t, the average profit achieved is

$$\mu_t \Pi_t^* = \phi\left(\frac{R_t}{A_t^*}\right)\Pi_t^*.$$

On the other hand, research activities are associated with costs R_t, and these costs must be borne regardless of whether they are successful. The average profit generated by research activities is therefore

$$\phi\left(\frac{R_t}{A_t^*}\right)\Pi_t^* - R_t.$$

In order to optimize this, the first-order condition equation is employed, i.e.,

$$0 = \frac{\partial\left(\phi\left(\frac{R_t}{A_t^*}\right)\Pi_t^* - R_t\right)}{\partial R_t} = \phi'(n_t)\frac{\Pi_t^*}{A_t^*} - 1,$$

and with (1.30) this leads to

$$\phi'(n) = \frac{1}{c_\Pi}. \tag{1.31}$$

Here, the time dependence of n has been suppressed, since the right-hand side of (1.31) does not depend on time. Using (1.29), (1.31) specializes to

$$n = (\lambda\sigma c_\Pi)^{\frac{1}{1-\sigma}},$$

and the probability of success μ of innovations, which is now independent of time, is

$$\mu = \lambda^{\frac{1}{1-\sigma}}(\sigma c_\Pi)^{\frac{\sigma}{1-\sigma}}. \tag{1.32}$$

In [4], a footnote points out that the parameters appearing in (1.32) should be chosen to be so small that the probability of success is $\mu < 1$.

Economic Growth and the Capital Growth Rate

All variables appearing in (1.28) are proportional to productivity A_t. Thus, the growth rates g_t of all these variables coincide and amount to

$$g_t = \frac{A_t - A_{t-1}}{A_{t-1}}.$$

Now, $A_t = \gamma A_{t-1}$ or $A_t = A_{t-1}$, depending on whether the innovation in period t was successful or not. Thus, $g_t = \gamma - 1$ in the event of success, which occurs with probability μ, and $g_t = 0$ otherwise. The expected growth rate g is therefore given by

$$g = \mu(\gamma - 1). \tag{1.33}$$

The probability μ can be interpreted as the long-term frequency of successful innovation efforts. Furthermore, $\gamma - 1$ is the increase in productivity resulting from each successful research activity. Therefore, the average growth rate g corresponds to the frequency of innovations multiplied by the increase in productivity.

Using (1.32), we can say that:

- Growth increases with research productivity λ.
- Growth increases with the incentive c_{Π} for innovation.
- Growth increases with population size L.

In particular, g in (1.33) denotes the growth of gross domestic product G_t, i.e., economic growth. If x_t is interpreted as capital K_t used for production, $K_t = x_t$, then the capital return rate r turns out to coincide with the economic growth rate g

$$r = \frac{K_t - K_{t-1}}{K_{t-1}} = g.$$

1.4 Classical Growth Theory and Wealth Concentration

The Schumpeterian model and the AK model imply that the rate of return on capital r corresponds to economic growth g, while the Solow-Swan model implies the inequality $r > g$—if $\dot{k}/k > 0$. In 2014, the French economist Thomas Piketty published *Capital in the Twenty-First Century* [17], summarizing his empirical research on wealth distribution across countries. He concluded that wealth concentration is observed if

$$r > g$$

holds in these countries.

This result sparked heavy criticism. For instance, economist N. Gregory Mankiw wrote in his article *Yes,$r > g$. So What?* [15]:

> The book [Piketty's Capital in the Twenty-First Century, JK] documents that the rate of return on private capital r exceeds the economy's growth rate g, and it argues that this will likely continue to be the case, perhaps by a larger amount in the future. Piketty (2014, p. 571) boldly calls this fact "the central contradiction of capitalism." He reasons that if $r > g$, the wealth of the capitalist class will grow faster than the incomes of workers, leading to an "endless inegalitarian spiral" (p. 572). To someone who views relatively unfettered capitalism as one of the great achievements of human history and the best way to organize a society, as I do, these conclusions present a significant challenge. The first thing to say about Piketty's logic is that it will seem strange to any economist trained in the neoclassical theory of economic growth. The condition $r > g$ should be familiar. In the textbook Solow growth model, it arrives naturally as a steady-state condition ...

Debraj Ray, an editor of *American Economic Review*, commented similarly in his article *Nit-Piketty—A comment on Thomas Piketty's Capital in the Twenty First Century* [18]:

> And so we come to Piketty's Third Fundamental Law, what he calls 'the central contradiction of capitalism':
> *The rate of return on capital systematically exceeds the overall rate of growth of income:* $r > g$.
> Relatively speaking, this is the most interesting of the three laws. It is a genuine prediction. It is falsifiable. And empirical research throws real light on this phenomenon: Piketty amasses data to argue that this inequality has held, by and large undisturbed, over a long period of time. (And reading this description of empirical trends, I continue to be impressed.)
> Here is what Piketty concludes from this Law, as do several approving reviewers of his book: that because the rate of return on capital is higher than the rate of growth overall, the income of capital owners must come to dominate as a share of overall income.
> ...
> But the Piketty faithful will still cling to the magic of that all-pervasive formula: $r > g$. That looks right, doesn't it, and besides, it is impressive how empirically the law appears to hold through decades of data. My answer is: yes, it does look right, and its empirical validity is indeed impressive, but to me it is impressive for a different reason: that it is a mini-triumph of economic *theory*.
> Here is a fact. Take any theory of economic growth that is fully compatible with 'balanced income growth' of all individuals, the kind we already know does a bad job in explaining rising inequalities. Under the mildest efficiency criterion—one that essentially states that it must be impossible to increase consumption for every generation, including the current generation, by lowering savings rates—it follows, not empirically but as a matter of theoretical prediction, that $r > g$. Piketty's Third Law has been known to economic theorists for at least 50 years, and no economic theorist has ever suggested that it 'explains' rising inequality. Because it doesn't. It can't, because the models that generate this finding are fully compatible with stable inequalities of income and wealth. (More on this in the Appendix.) You need something else to get at rising inequality.

Both Mankiw and Ray argue that $r > g$ cannot explain growing inequality, since standard growth theory already implies this relationship and no economist has ever claimed that $r > g$ could "explain" rising inequality.

In what follows, this view is challenged: We show that a link between $r > g$ and increasing inequality can be established within growth-theoretic modeling. This requires examining not only the economy as a whole but also the heterogeneous household structures that comprise it. The economic trajectory of wealthier groups can differ significantly from that of less wealthy groups. In particular, the initial endowment—labor income, consumption, and capital—and the relationship between r and g fundamentally determine each group's development path.

Chapter 2
The Circular Flow Model Without State

For decades, growth theory has sought to explain growth endogenously, i.e., from within the respective model.

This chapter marks a departure from this paradigm. In the following, we will not base our analysis on physical capital but will instead examine how the wealth of an economy as a whole and its household groups develop over time. Wealth is interpreted in monetary terms and includes both financial and real assets. In the modeling we will use capital return rates that are specified exogenously.

In addition, the conditions that influence the growth of an economy are considered so complex that no attempt will be made to describe them functionally. Not only research, innovation, technical progress, sufficient and affordable energy supplies, a functioning healthcare system, an efficient education system, a reliable legal framework, functioning state institutions, and an intellectual atmosphere that promotes creativity and productivity but also a stable peace order and functioning international trade in raw materials essential for production are complex factors that cannot be captured in simple formulas. Therefore, economic growth is also specified exogenously—rather than endogenously—in the following.

However, further analysis is then based on generally applicable balance equations that will be derived from circular flow models. The idea of circular flow models first appeared explicitly in the *Tableaux Economiques* of the physiocrats, which were developed by their most important representative, François Quesnay, in the eighteenth century. The circular flow model for the overall economy will turn out to be a modified AK model in Sect. 3.2, and thus the circular flow models presented in this book are linked to one of the fundamental models of growth theory specified in [4].

The most important result of the extensive empirical research conducted by economist Thomas Piketty, which he published in his book *Capital in the 21st Century*, [17], is that he found a correlation between increasing economic inequality

J. Kremer, *Capital Income and Concentration of Wealth*, Springer Studies in Alternative Economics, https://doi.org/10.1007/978-3-658-50715-2_2

and capital concentration. This connection cannot be explained by standard growth models, but it does follow from an analysis of circular flow models, as we shall see.

In Sect. 2.1, the circular flow model for an economy without state is presented, and balance equations for various economic variables are derived. The model is extended in Sect. 2.2 to account for real assets. In Sect. 2.3, heterogeneous household groups and dynamic behavior over time are introduced.

Consumer spending by households covers all costs of the firms that supply goods and services, including capital costs. Conversely, households that own capital can earn capital income. The difference between capital income and capital costs is a transfer payment called capital income transfer or simply interest transfer. This mechanism causes a redistribution of wealth toward the wealthy, as shown in Sect. 2.4 through a simple example.

2.1 Balance Equations of the Circular Flow Model Without State

The model considered here is shown in Fig. 2.1 and initially includes only financial capital. The nodes labeled *households* and *firms* represent the respective economic agents. We analyze the monetary inflows and outflows at each node and derive accounting equations that form the core of the model.

In Chap. 8, we will expand the model to include the state as an economic agent and take the fiscal system into account.

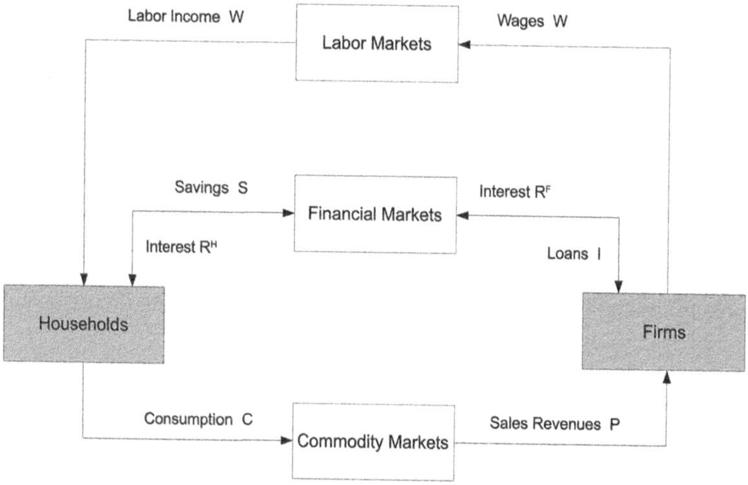

Fig. 2.1 Circular flow model without state

Households

Households receive income from two sources: **wages** W and **interest income** R^H. Wages come from labor and are therefore also called **labor income**. In contrast, R^H represents capital income, derived from ownership of **wealth** V, assumed for now to be monetary. Total household income, $W + R^H$, is used for **consumption** C and **savings** S, thus

$$W + R^H = C + S. \tag{2.1}$$

If S is positive, then wealth V increases by the amount saved. If S is negative, V decreases. Even when aggregate savings are positive, individual households may reduce their wealth or take out loans.

Firms

Firm revenues consist of **sales** P and **borrowed funds** I, also called **loans**. Expenses include **wages** W and a share of **financing costs** R^F, which represents the portion passed on by banks to depositors. The remaining financing costs are retained by banks and counted as part of wages W. Thus we have

$$P + I = W + R^F. \tag{2.2}$$

If I is positive, it indicates new borrowing, and if is negative, it reflects savings or loan repayments.

Macroeconomic Relations Between Households and Firms

From the circular flow model we derive that sales by firms equal household consumption

$$P = C. \tag{2.3}$$

In our financial system, increases in aggregate balances result from credit expansion and decreases from amortization or write-offs. This translates to

$$S = I. \tag{2.4}$$

That is, any change in monetary wealth V is mirrored by an equal change in debt. Hence, aggregate monetary wealth equals total **debt** B in the economy.

If savings S occur, then part of W is not spent on C, and firms must borrow $I = S$. Alternatively, firms may invest, borrowing amount I, which then appears as S in depositors' accounts.[1]

Further, we have

$$R^H = R^F = R. \tag{2.5}$$

The amount R^F is the share of the financing costs that is handed over to depositors as interest revenues R^H.[2] The remaining share of the financing costs paid by borrowers results in labor income of the banks and is thus part of W. R^H and R^F coincide by definition and are thus abbreviated by the symbol R. From (2.1), (2.2), (2.3), (2.4), and (2.5), we obtain

$$\begin{array}{ccccccc} W & + & R^H & = & C & + & S \\ \| & & \| & & \| & & \| \\ W & + & R^F & = & P & + & I. \end{array} \tag{2.6}$$

Gross domestic product (GDP) Y is defined as

$$Y = W + R = P + I = C + S. \tag{2.7}$$

This can be viewed as total firm expenditures, total firm revenues, total household income, or total household expenditures. Economic growth is defined as growth in Y.

A particularly insightful identity is

$$Y = W + R. \tag{2.8}$$

If Y remains constant while R rises (e.g., due to compound interest), labor income W must fall, and capital owners receive an increasing share of Y. Conversely, if Y grows fast enough, W can rise or remain stable despite increasing R. Thus, if there are no limits to the growth of capital income, this exerts pressure on the economy to grow.

---

[1] The textbook identity *savings* = *investments* differs from $S = I$. Standard economics assumes saved deposits are lent to borrowers, but in our monetary systems, deposits and loans are created simultaneously.

[2] This does not imply that interest rates for loans and savings coincide.

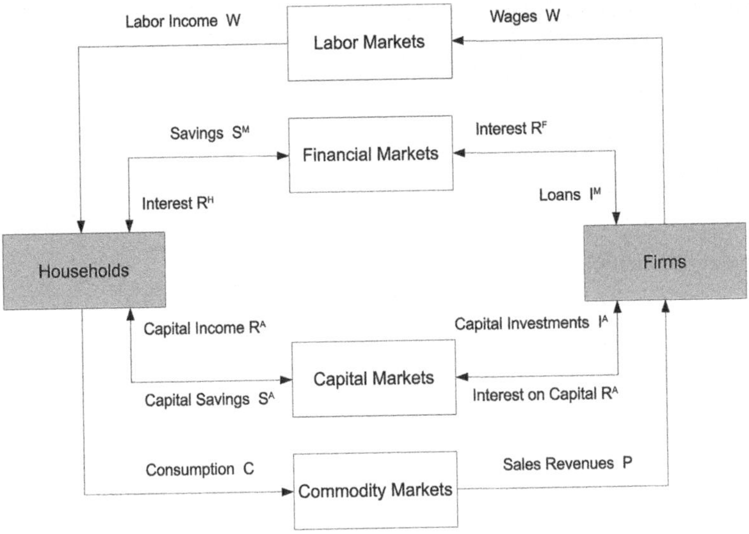

Fig. 2.2 Circular flow model without government, including monetary and real assets

2.2 Consideration of Real Assets

In addition to acquiring real assets for consumption, households can also invest in profitable real capital. This includes the purchase of properties and real estate for rental or leasing purposes. Furthermore, money can be invested in stocks to acquire shares in companies. When these companies pay dividends, investors participate in company profits. In all such cases, the revenues—rents, lease payments, and dividends—are forms of capital income, analogous to interest income from monetary wealth.

Figure 2.2 shows an extension of the circular flow model that incorporates real asset investments. Monetary savings are denoted by S^M, real asset savings by S^A, and capital income from real assets by R^A.

If households purchase newly issued stocks, the payment S^A becomes a cash inflow I^A for the corresponding companies. Dividends received in return are recorded as capital income R^A. However, if one household buys stocks from another, then neither S^A nor S^M changes—the transaction is a redistribution between households.

Households that rent or lease out real estate are modeled as firms. The capital savings S^A are considered investments I^A into these firms. The households that use these properties pay rents or leases as part of their consumption C to the provider firms.

The landlords initially use these payments for maintenance and modernization, resulting in wage payments to other firms. The remaining amount is divided between the landlord's labor income W for services rendered and capital income R^A from ownership of tangible assets.

Thus, the household balance equation extends to

$$W + R^H + R^A = C + S^M + S^A, \tag{2.9}$$

and for firms we now have

$$W + R^F + R^A = P + I^M + I^A. \tag{2.10}$$

We also retain the relations

$$\begin{aligned} C &= P, \\ R^H &= R^F, \\ S^M &= I^M, \\ S^A &= I^A. \end{aligned}$$

Defining

$$\begin{aligned} S &= S^M + S^A, \\ I &= I^M + I^A, \\ R &= R^H + R^A, \end{aligned}$$

we obtain

$$S = I$$

and the updated macroeconomic identity

$$W + R = C + S = P + I = Y. \tag{2.11}$$

This matches the expression in (2.7), but now R denotes **capital income** from both monetary and real assets. The relationships in (2.11) are illustrated in Fig. 2.3, where the financial and real capital markets are merged into a single **capital markets** node.

Again, the expressions in (2.11) are referred to as **gross domestic product** Y. This representation corresponds to Eq. (2.7) from Sect. 2.1.

However, the equivalence of aggregate wealth and aggregate debt only applies to monetary wealth. Real assets are not offset by liabilities. If we denote financial wealth by V^M, real capital wealth by V^A, and **total wealth** by V, we obtain

$$V = V^M + V^A,$$

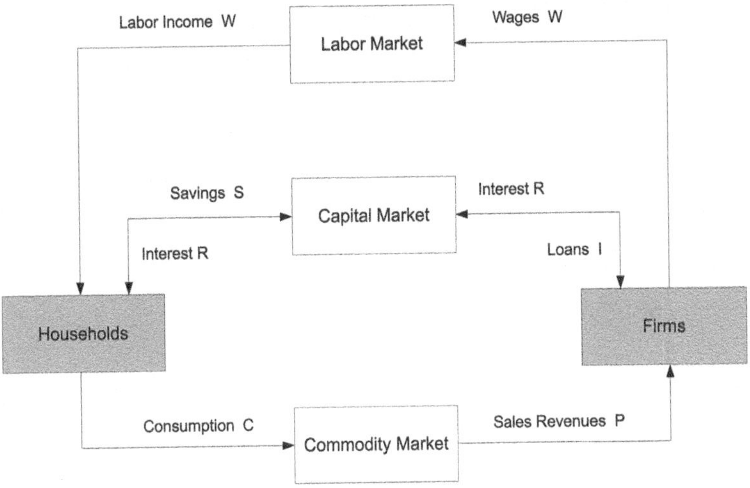

Fig. 2.3 Circular flow model without government, with capital markets

but

$$V^M = B,$$

if B denotes **total debt**.

2.3 The Circular Flow Model Without State

The main goal of the models presented in this book is to examine heterogeneous economies. To do this, the economy under consideration is divided into N household groups, each comprising the same number of households that share identical economic characteristics such as wealth, consumption, savings, and income. Macroeconomic variables specific to household groups are denoted with superscripts. Thus, W^i represents the labor income of group i, where $i = 1, \ldots, N$. The corresponding symbol without upper index denotes the aggregated value, so that $W = \sum_{i=1}^N W^i$.

We consider the development of the economy over a time period from today, $t = 0$, until some future time $T > 0$. All quantities are indexed by time using a subscript. For instance, W_t^i denotes the labor income of group i at time t, and $W_t = \sum_{i=1}^N W_t^i$.

We model **economic growth** with growth factors g_t for $t \geq 0$ so that

$$\dot{Y}_t = g_t Y_t, \tag{2.12}$$

where Y_t denotes the GDP for time t. From (2.11), we have

$$Y_t = W_t + R_t = C_t + S_t. \quad (2.13)$$

We further assume that overall consumption changes at rates c_t so that

$$\dot{C}_t = c_t C_t. \quad (2.14)$$

Solving the differential equations (2.12) and (2.14) for Y and C leads to

$$S_t = Y_t - C_t \quad (2.15)$$

and thus to

$$V_t = V_0 + \int_0^t S_s \, ds. \quad (2.16)$$

Let r_t denote the **capital return rate** at time t, then

$$R_t = r_t V_t. \quad (2.17)$$

Labor income at time t can now be expressed as

$$W_t = Y_t - r_t V_t. \quad (2.18)$$

To distribute W_t among the household groups, we define factors $\alpha_t^i \geq 0$ such that $\sum_{i=1}^N \alpha_t^i = 1$ and

$$W_t^i = \alpha_t^i W_t. \quad (2.19)$$

This ensures $\sum_{i=1}^N W_t^i = \left(\sum_{i=1}^N \alpha_t^i\right) W_t = W_t$. Each group also contributes to total consumption C_t. Defining factors $\beta_t^i > 0$ with $\sum_{i=1}^N \beta_t^i = 1$, we write

$$C_t^i = \beta_t^i C_t \quad (2.20)$$

so that $\sum_{i=1}^N C_t^i = C_t$. For each group $i = 1, \ldots, N$, we assume initial wealth V_0^i is given at $t = 0$. Wealth evolves as

$$V_t^i = V_0^i + \int_0^t S_s^i \, ds \quad (2.21)$$

so that

$$\dot{V}_t^i = S_t^i. \quad (2.22)$$

Substituting

$$S_t^i = W_t^i + R_t^i - C_t^i$$

and

$$R_t^i = r_t V_t^i$$

into (2.22) leads to

$$\dot{V}_t^i = \alpha_t^i W_t + r_t V_t^i - \beta_t^i C_t. \tag{2.23}$$

These differential equations determine the evolution of each group's wealth over time.

In addition to initial values, the model requires specification of the following functions for $0 \leq t \leq T$:

1. Capital return rates r_t
2. Growth rates g_t of GDP Y_t
3. Growth rates c_t of consumption C_t
4. Distribution factors α_t^i for labor income W_t
5. Distribution factors β_t^i for consumption C_t

Once these inputs are defined, the model is fully specified. Such an economy is referred to as a **circular flow economy**.

Chapter 3 will show that if all parameters are constant over time, the system's behavior can be derived analytically. If parameters vary with time, the development of the economic variables can be computed recursively using a discretized version of the model. This case will be explored in the second part of the book.

2.4 Interest Transfer

This section introduces a redistribution mechanism called interest transfer. Every member of an economy pays capital costs and may receive returns on invested capital. Importantly, capital costs must be borne even without taking out a loan. This is because capital costs are embedded in consumer spending and taxation: Companies calculate the prices of their goods and services to cover all their costs, including financing costs.

Dividing total capital income R_t in proportion to each household group's consumption yields the capital cost

$$\Pi_t^i = \frac{C_t^i}{C_t} R_t = \beta_t^i R_t$$

hidden in consumption for group i. The balance Z_t^i between the interest income R_t^i and the interest cost Π_t^i for group i is thus given by

$$Z_t^i = R_t^i - \Pi_t^i. \tag{2.24}$$

We call Z_t^i the **interest transfer** of group i at time t. Since

$$\sum_{i=1}^{N} Z_t^i = \sum_{i=1}^{N} R_t^i - \frac{R_t}{C_t} \sum_{i=1}^{N} C_t^i = 0,$$

the positive balances of some groups must be offset by negative balances of others.

Let us illustrate this with an example of an economy composed of three household groups, with the following initial values:

Economy with three household groups

Group	Consumption	Wealth
1	10	0
2	30	100
3	60	500

The total wealth of the economy is $V = 600$. Assuming a capital return rate $r = 5\%$, the total capital income is

$$R = r \cdot V = 5\% \cdot 600 = 30.$$

This capital income—including interest, dividends, rents, and leases—is funded via consumer prices and thus through household spending. The total consumption is

$$C = 10 + 30 + 50 = 90.$$

Now we compute each group's capital cost in consumption:

Capital cost

Group	Consumption share	Capital cost Π^i
1	$\frac{10}{100} = 10\%$	$30 \cdot 10\% = 3$
2	$\frac{30}{100} = 30\%$	$30 \cdot 30\% = 9$
3	$\frac{60}{100} = 60\%$	$30 \cdot 60\% = 18$

Next, we compute capital income $R^i = rV^i$ for each group:

Capital income

Group	Wealth	Capital income R^i
1	0	$0 \cdot 5\% = 0$
2	100	$100 \cdot 5\% = 5$
3	500	$500 \cdot 5\% = 25$

Finally, interest transfer $Z^i = R^i - \Pi^i$ is calculated:

Interest transfer

Group	Capital income – Capital cost
1	$0 - 3 = -3$
2	$5 - 9 = -4$
3	$25 - 18 = 7$

We observe that the wealthiest group (Group 3) has a positive balance: Its investment income exceeds its capital costs, despite high consumption. The resulting surplus is a transfer payment—financed entirely by Groups 1 and 2.

This illustrates the redistribution of wealth "from the bottom to the top" due to capital returns. In the next chapter, we will show that interest transfer is the main driver of redistribution and wealth concentration.

Chapter 3
The Circular Flow Model with Constant Parameters

In this chapter, we explore the circular flow model from Sect. 2.3 under the assumption of time-constant parameters. In this special case, closed-form solutions for the evolution of important macroeconomic variables—such as labor income and wealth—for different household groups can be derived.

We assume $S_0 > 0$, $C_0 > 0$, and $V_0 > 0$, along with a constant capital return rate $r_t = r$ and a constant economic growth rate $g_t = g$, so that

$$Y_t = e^{gt} Y_0 \tag{3.1}$$

with $Y_0 = S_0 + C_0 > 0$. We further assume growth in total consumption at a rate $c_t = c$ and conclude

$$C_t = e^{ct} C_0.$$

Thus we find

$$S_t = Y_t - C_t = e^{gt} Y_0 - e^{ct} C_0.$$

This implies the asymptotic behavior of the savings rate S_t / Y_t as $t \to \infty$:

$$\frac{S_t}{Y_t} = 1 - e^{(c-g)t} \frac{C_0}{Y_0} \to \begin{cases} 1 & (c < g) \\ S_0/Y_0 & (c = g) \\ -\infty & (c > g)\,. \end{cases}$$

To avoid unrealistic outcomes, we set $c = g$, so that

$$C_t = e^{gt} C_0. \tag{3.2}$$

J. Kremer, *Capital Income and Concentration of Wealth*, Springer Studies in Alternative Economics, https://doi.org/10.1007/978-3-658-50715-2_3

In addition, for labor income and consumption we assume time-independent shares $0 \leq \alpha^i \leq 1$, $\sum_{i=1}^N \alpha^i = 1$, and $0 \leq \beta^i \leq 1$, $\sum_{i=1}^N \beta^i = 1$, so that

$$\begin{aligned} W_t^i &= \alpha^i W_t, \\ C_t^i &= \beta^i C_t \end{aligned} \tag{3.3}$$

for $i = 1, \ldots, N$ and for all $t \geq 0$.

3.1 Evolution of the Economy

Lemma 3.1 *For aggregated savings of the households, we have*

$$S_t = e^{gt} S_0, \tag{3.4}$$

and the total wealth of the economy is given by

$$V_t = \begin{cases} t S_0 + V_0 & (g = 0) \\ e^{gt} \frac{1}{g} S_0 + \left(V_0 - \frac{1}{g} S_0\right) & (g \neq 0). \end{cases} \tag{3.5}$$

Proof Equation (3.4) follows with (3.1) and (3.2) from $S_t = Y_t - C_t$. With (2.16) this leads to

$$V_t = V_0 + S_0 \int_0^t e^{gs}\, \mathrm{d}s = \begin{cases} V_0 + t S_0 & (g = 0) \\ V_0 + \frac{e^{gt} - 1}{g} S_0 & (g \neq 0). \end{cases}$$

□

Thus, the growth of total wealth depends on economic growth g, but not on the capital return rate r. Figure 3.1 shows the development of total wealth over time for two parameter sets.

Corollary 3.2 *With $V_0 > 0$ and $S_0 > 0$, the growth rate $\dot{V}_t / V_t$ of V_t is asymptotically given by*

$$\lim_{t \to \infty} \frac{\dot{V}_t}{V_t} = \begin{cases} g & (g > 0) \\ 0 & (-1 < g \leq 0). \end{cases}$$

Furthermore, $\dot{V}_t / V_t > 0$ for t large enough.

Proof From (3.5), we conclude

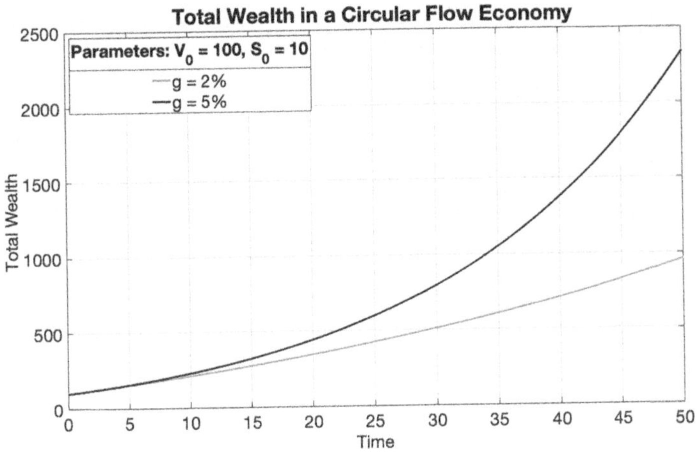

Fig. 3.1 Total wealth V_t in a circular flow economy for two parameter sets

$$\frac{\dot{V}_t}{V_t} = \begin{cases} \frac{S_0}{V_0 + t S_0} & (g = 0) \\ g \frac{S_0}{S_0 + e^{-gt}(g V_0 - S_0)} & (g \neq 0) \, . \end{cases}$$

Case $g = 0$. Here, we have $S_0/(V_0 + t S_0) > 0$ for all $t \geq 0$ and $\lim_{t\to\infty} \dot{V}_t / V_t = 0$.

Now, assume $g \neq 0$, and write

$$g \frac{S_0}{S_0 + e^{-gt}\,(g V_0 - S_0)} = \frac{g}{1 + a e^{-gt}}$$

with $a = g V_0 / S_0 - 1$.

Case $g > 0$. In this case, we have $\lim_{t\to\infty} e^{-gt} = 0$ so that

$$\lim_{t\to\infty} \frac{g}{1 + a e^{-gt}} = g$$

and $g/\left(1 + a e^{-gt}\right) > 0$ for t large enough.

Case $g < 0$. Now we have $a < 0$, so that $1 + a e^{-gt} \to -\infty$ for $t \to \infty$. Thus, as in the previous case we have $g/\left(1 + a e^{-gt}\right) > 0$ for t large enough, but

$$\lim_{t\to\infty} \frac{g}{1 + a e^{-gt}} = 0.$$

□

According to Corollary 3.2, the growth rate of total wealth V_t is approximately given by the economic growth rate g for large times if $g > 0$.

Lemma 3.3 *In a circular flow economy with constant parameters, the total wage is given by*

$$W_t = \begin{cases} -tr\,S_0 + W_0 & (g = 0) \\ e^{gt}\left(Y_0 - \frac{r}{g}S_0\right) + \left(\frac{r}{g}S_0 - r\,V_0\right) & (g \neq 0)\,. \end{cases} \tag{3.6}$$

Proof For $g = 0$ we find with (3.1) and (3.2)

$$Y_t = Y_0, \quad C_t = C_0.$$

Equations (3.4) and (3.5) imply $S_t = Y_0 - C_0 = S_0$ and

$$V_t = V_0 + t\,S_0.$$

Due to $R_t = r\,V_t$ and (2.13), we conclude

$$\begin{aligned} W_t &= Y_t - r\,V_t \\ &= Y_0 - R_0 - tr\,S_0 \\ &= W_0 - tr\,S_0. \end{aligned}$$

For $g \neq 0$ we find with (3.5)

$$\begin{aligned} W_t &= Y_t - r\,V_t \\ &= e^{gt}\left(Y_0 - \frac{r}{g}S_0\right) + \left(\frac{r}{g}S_0 - r\,V_0\right), \end{aligned}$$

as was to be shown. □

Thus, total labor income W_t depends not only on economic growth g but also on the capital return rate r. From Lemma 3.3 we conclude that for $g > 0$ and $r > 0$ total labor income W_t will increase exponentially in the case of $Y_0 > \frac{r}{g}S_0$. But if $Y_0 < \frac{r}{g}S_0$, the total labor income will decrease to and below zero, although for $g > 0$ total wealth will increase exponentially. Thus, if economic growth is not strong enough, this already indicates bleak prospects for the economic development of household groups with low wealth. Finally, if

$$Y_0 = \frac{r}{g}S_0, \tag{3.7}$$

then labor income remains constant over time.

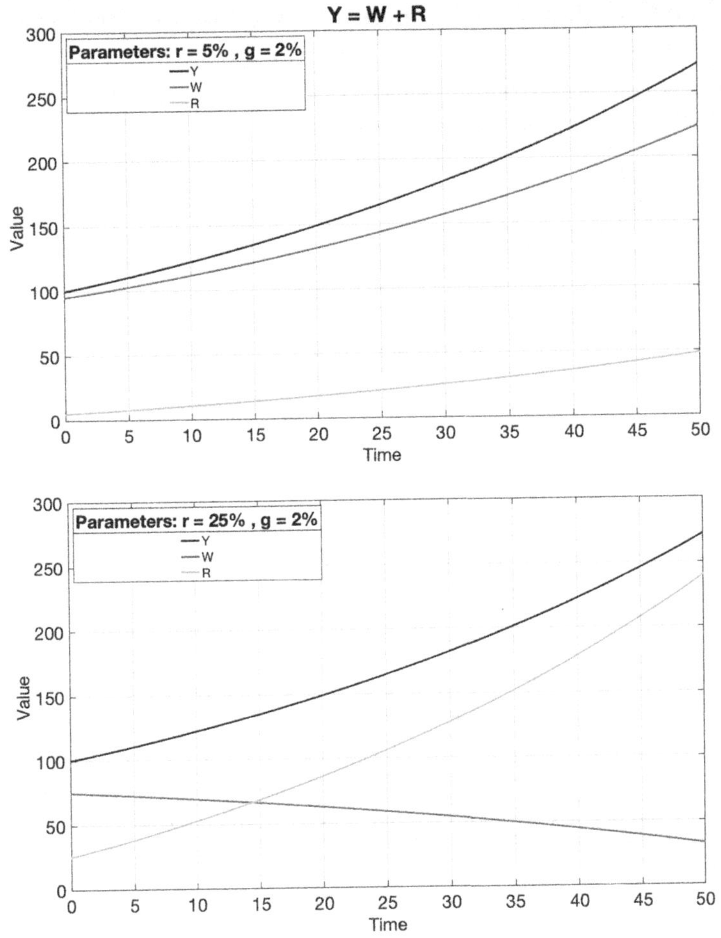

Fig. 3.2 Decomposition of the gross domestic product $Y_t = W_t + R_t$ of the economy into labor income W_t and capital income R_t for two parameter sets. In the computations, $Y_0 = V_0 = 100$, and $S_0 = 10$ was assumed in both cases

Figure 3.2 shows the development of the economy's gross domestic product Y_t over time and its decomposition into labor income W_t and capital income R_t for two parameter sets. The upper part of Fig. 3.2 shows the decomposition for $r = 5\%$ and $g = 2\%$ and the lower part the decomposition for $r = 25\%$ and $g = 2\%$. Since Y_t only depends on g, but not on r, GDP develops identically in both cases. What is different is the allocation of Y_t to labor income and investment income. In the latter case, the rate of return on capital is so high that aggregate labor income declines over time, while capital income accounts for an ever larger share of GDP.

3.2 The Circular Flow Economy as AK Model with Offset

The evolution (3.1) of the economy may be formulated as a modified AK model. To this end we assume $g \neq 0$ and use (3.5) to write

$$\begin{aligned} Y_t &= e^{gt} Y_0 \\ &= \frac{gY_0}{S_0}\left(\frac{S_0}{g}e^{gt} + \left(V_0 - \frac{S_0}{g}\right)\right) - \frac{gY_0}{S_0}\left(V_0 - \frac{S_0}{g}\right) \\ &= AV_t + B \end{aligned} \tag{3.8}$$

with

$$\begin{aligned} A &= \frac{gY_0}{S_0}, \\ B &= Y_0 - AV_0. \end{aligned} \tag{3.9}$$

If we identify V with the capital stock and Y with output and if we interpret

$$F(V) = AV + B$$

as a production function, we write (3.8) as

$$Y_t = F(V_t), \tag{3.10}$$

and the circular flow model for the total economy may be expressed as an **AK model with offset** B.

According to (3.1) and (3.4), the savings rate is

$$s = \frac{S_t}{Y_t} = \frac{S_0}{Y_0}, \tag{3.11}$$

which is constant over time. This and (2.16) lead to

$$\begin{aligned} \dot{V}_t &= S_t \\ &= sY_t \\ &= sAV_t + sB \\ &= gV_t + (S_0 - gV_0). \end{aligned}$$

A special solution V_t^1 of the linear inhomogeneous differential equation

$$\dot{V}_t - gV_t = S_0 - gV_0 \tag{3.12}$$

is given by the constant function

$$V_t^1 = V_0 - \frac{1}{g} S_0.$$

The general solution V_t^2 of the corresponding homogeneous equation

$$\dot{V}_t - g V_t = 0$$

is given by

$$V_t^2 = c e^{gt},$$

where c denotes an arbitrary constant. The general solution V_t of (3.12) may thus be written as

$$V_t = V_t^1 + V_t^2.$$

If a value V_0 is specified for $t = 0$, we obtain the solution

$$V_t = e^{gt} \frac{1}{g} S_0 + \left(V_0 - \frac{1}{g} S_0 \right), \tag{3.13}$$

and this coincides with (3.5). Using (3.11) and (3.13), we finally get

$$Y_t = \frac{1}{s} S_t = \frac{Y_0}{S_0} \dot{V}_t = Y_0 e^{gt}. \tag{3.14}$$

Capital Return Rate for Constant Labor Income

According to (3.13) and (3.14), V_t and Y_t are independent of r, but both capital income and labor income depend on r:

$$R_t = r V_t, \quad W_t = Y_t - R_t = e^{gt} \left(Y_0 - \frac{r}{g} S_0 \right) - \left(r V_0 - \frac{r}{g} S_0 \right).$$

In the given circular flow model, economic growth g and the rate of return on capital r can be chosen independently of each other, and the rate of return on capital cannot be inferred from the model for the overall economy.

The return rate $\bar{r}$ that leads to constant W_t is

$$\bar{r} = \frac{Y_0}{S_0} g > g.$$

If $r > \bar{r}$, W_t falls below zero, risking economic collapse. If $r < \bar{r}$, W_t rises indefinitely.

However, the next section shows the broader consequences of $r > g$ when considering household group dynamics.

3.3 Evolution of the Household Groups

This section analyzes the temporal development of key economic variables—particularly wealth and labor income—for different household groups.

We consider different parameter constellations for the capital return rate r and the economic growth rate g:

$$\begin{array}{lll} r = 0 & g = 0 & \\ r = 0 & g \neq 0 & \\ r \neq 0 & g = 0 & \\ r \neq 0 & g \neq 0 & r = g \\ r \neq 0 & g \neq 0 & r \neq g . \end{array}$$

According to Sect. 2.3, we have

$$\dot{V}_t^i = S_t^i \tag{3.15}$$

and

$$\begin{aligned} S_t^i &= W_t^i + R_t^i - C_t^i \\ &= \alpha^i W_t + r V_t^i - \beta^i C_t \end{aligned} \tag{3.16}$$

for $i = 1, \ldots, N$. Using (3.6) and (3.2), we obtain

$$\dot{V}_t^i = \begin{cases} r V_t^i - t\alpha^i r S_0 + \alpha^i W_0 - \beta^i C_0 & (g = 0) \\ r V_t^i + e^{gt}\left(\alpha^i \left(Y_0 - \frac{r}{g} S_0\right) - \beta^i C_0\right) + \alpha^i \left(\frac{r}{g} S_0 - r V_0\right) & (g \neq 0) . \end{cases} \tag{3.17}$$

Thus, the ordinary linear inhomogeneous first-order differential equations specified in (3.17) determine the wealth development of the household groups.

Lemma 3.4 *In a circular flow economy with constant parameters, the wealth V_t^i of the ith household group is for $r = 0$ given by*

$$V_t^i = \begin{cases} t S_0^i + V_0^i & (g = 0) \\ e^{gt} \frac{1}{g} S_0^i + \left(V_0^i - \frac{S_0^i}{g}\right) & (g \neq 0) . \end{cases} \tag{3.18}$$

Proof For $r = 0$ and $g = 0$, the first line of (3.17) is

$$\dot{V}_t^i = W_0^i - C_0^i = S_0^i,$$

where (3.16) was used for the second equality. Integration gives the first line of (3.18).

For $r = 0$ and $g \neq 0$, the second line of (3.17) simplifies to

$$\dot{V}_t^i = e^{gt}\left(\alpha^i Y_0 - \beta^i C_0\right) = e^{gt} S_0^i,$$

using $Y_0 = W_0 + rV_0 = W_0$. Integration leads to the second line of (3.18). □

In this section we will utilize the following proposition:

Proposition 3.5 *Let $I \subset \mathbb{R}$ be an open interval with $0 \in I$, and let $f, g : I \to \mathbb{R}$ be continuous functions. Then, for each $x_0 \in \mathbb{R}$, there is a uniquely determined solution of the initial value problem*

$$\dot{x}(t) = f(t)\, x(t) + g(t), \quad x(0) = x_0. \tag{3.19}$$

The solution is given by

$$x(t) = e^{F(t)}\left(x_0 + \int_0^t e^{-F(s)} g(s)\, \mathrm{d}s\right) \tag{3.20}$$

with

$$F(t) = \int_0^t f(s)\, \mathrm{d}s.$$

Proof To check that x in (3.20) solves the initial value problem (3.19), we calculate the time derivative of x and obtain

$$\begin{aligned}\dot{x}(t) &= \dot{F}(t)\, e^{F(t)}\left(x_0 + \int_0^t e^{-F(s)} g(s)\, \mathrm{d}s\right) + e^{F(t)} e^{-F(t)} g(t) \\ &= f(t)\, x(t) + g(t).\end{aligned}$$

An elegant proof of the uniqueness of (3.20) can be found in [1]. □

The following abbreviations will occur repeatedly:

$$\begin{aligned}\Delta_1^i &= \left(\alpha^i - \beta^i\right) C_0, \\ \Delta_2^i &= V_0^i - \alpha^i V_0\end{aligned} \tag{3.21}$$

$$= \left(\upsilon^i - \alpha^i\right) V_0$$

with

$$\upsilon^i = \frac{V_0^i}{V_0}. \tag{3.22}$$

At the beginning of this chapter, $V_0 > 0$ was assumed, so (3.22) is well defined. In the next section, we will see that $\Delta_1^i > 0$ and $\Delta_2^i > 0$ characterize wealthy household groups, and $\Delta_1^i < 0$ and $\Delta_2^i < 0$ indicate less wealthy ones.

Lemma 3.6 *For $r \neq 0$ and $g = 0$, the wealth V_t^i of the ith household group is given by*

$$V_t^i = e^{rt}\left(\frac{1}{r}\Delta_1^i + \Delta_2^i\right) + t\alpha^i S_0 + \left(V_0^i - \left(\frac{1}{r}\Delta_1^i + \Delta_2^i\right)\right). \tag{3.23}$$

Proof For $r \neq 0$ and $g = 0$, we have according to (3.17)

$$\dot{V}_t^i = rV_t^i - tr\alpha^i S_0 + \alpha^i W_0 - \beta^i C_0.$$

This differential equation has the structure

$$\dot{x} = rx + at + b$$

with

$$x = V^i, \quad a = -r\alpha^i S_0, \quad b = W_0^i - C_0^i. \tag{3.24}$$

We apply Proposition 3.5 with $f(t) = r$ and $g(t) = at + b$ and compute with $F(t) = rt$

$$\begin{aligned} x(t) &= e^{rt}\left(x_0 + \int_0^t e^{-rs}(as + b)\, \mathrm{d}s\right) \\ &= e^{rt}\left(x_0 + \frac{a}{r^2} + \frac{b}{r}\right) - a\frac{t}{r} - \frac{a}{r^2} - \frac{b}{r}. \end{aligned}$$

Next we substitute (3.24) and obtain

$$V_t^i = e^{rt}\left(V_0^i + \frac{1}{r}\left(W_0^i - \alpha^i S_0 - C_0^i\right)\right) + \alpha^i S_0 t - \frac{1}{r}\left(W_0^i - \alpha^i S_0 - C_0^i\right).$$

The computation

$$\begin{aligned}
\alpha^i W_0 - \alpha^i S_0 - \beta^i C_0 &= \alpha^i (W_0 - S_0) - \beta^i C_0 \\
&= \alpha^i (C_0 - R_0) - \beta^i C_0 \\
&= \Delta_1^i - \alpha^i r V_0 \\
&= \Delta_1^i + r\Delta_2^i - r V_0^i
\end{aligned}$$

finally leads to (3.23). □

Now we consider the case $g \neq 0$. With (3.17) and $Y_0 = C_0 + S_0$, we find

$$\dot{V}_t^i = r V_t^i + e^{gt} \left(\left(\alpha^i - \beta^i \right) C_0 + \alpha^i S_0 \left(1 - \frac{r}{g} \right) \right) + \alpha^i r \left(\frac{1}{g} S_0 - V_0 \right).$$

This differential equation is of type

$$\dot{x} = rx + ae^{gt} + b \tag{3.25}$$

with

$$x = V^i, \quad a = \left(\alpha^i - \beta^i \right) C_0 + \alpha^i S_0 \left(1 - \frac{r}{g} \right), \quad b = \alpha^i r \left(\frac{1}{g} S_0 - V_0 \right). \tag{3.26}$$

If we apply Proposition 3.5 to (3.25) with $f(t) = r$ and $g(t) = ae^{gt} + b$, we obtain with $F(t) = rt$

$$\begin{aligned}
x(t) &= e^{rt} \left(x_0 + \int_0^t e^{-rs} \left(ae^{gs} + b \right) \mathrm{d}s \right) &(3.27)\\
&= e^{rt} \left(x_0 + a \int_0^t e^{(g-r)s} \, \mathrm{d}s + b \int_0^t e^{-rs} \, \mathrm{d}s \right).
\end{aligned}$$

The case $r = 0$ has already been covered in (3.18). Thus, for the evaluation of (3.27), we have to distinguish the cases

$$r \neq 0, \ g \neq 0, \ r \neq g$$

and

$$r \neq 0, \ g \neq 0, \ r = g.$$

Lemma 3.7 *For $r \neq 0$, $g \neq 0$, and $g \neq r$, the wealth V_t^i of the ith household group is given by*

$$V_t^i = e^{rt}\left(\frac{1}{r-g}\Delta_1^i + \Delta_2^i\right) + e^{gt}\left(\frac{1}{g-r}\Delta_1^i + \frac{\alpha^i}{g}S_0\right) + \alpha^i\left(V_0 - \frac{S_0}{g}\right). \tag{3.28}$$

Proof For $r \neq 0$ and $g \neq 0$ with $g \neq r$, (3.27) becomes

$$\begin{aligned} x(t) &= e^{rt}\left(x_0 + a\int_0^t e^{(g-r)s}\,\mathrm{d}s + b\int_0^t e^{-rs}\,\mathrm{d}s\right) \\ &= e^{gt}\frac{a}{g-r} + e^{rt}\left(\frac{a}{r-g} + x_0 + \frac{b}{r}\right) - \frac{b}{r}. \end{aligned}$$

Using (3.26), this can be written as

$$\begin{aligned} V_t^i = {} & e^{rt}\left(\frac{\left(\alpha^i - \beta^i\right)C_0 + \alpha^i S_0\left(1 - \frac{r}{g}\right)}{r-g} + \alpha^i\left(\frac{1}{g}S_0 - V_0\right) + V_0^i\right) \\ & + e^{gt}\frac{\left(\alpha^i - \beta^i\right)C_0 + \alpha^i S_0\left(1 - \frac{r}{g}\right)}{g-r} + \alpha^i\left(V_0 - \frac{S_0}{g}\right). \end{aligned}$$

We compute

$$\begin{aligned} \frac{\left(\alpha^i - \beta^i\right)C_0 + \alpha^i S_0\left(1 - \frac{r}{g}\right)}{r-g} + \alpha^i\left(\frac{1}{g}S_0 - V_0\right) + V_0^i &= \frac{1}{r-g}\Delta_1^i + V_0^i - \alpha^i V_0 \\ &= \frac{1}{r-g}\Delta_1^i + \Delta_2^i \end{aligned}$$

and

$$\frac{\left(\alpha^i - \beta^i\right)C_0 + \alpha^i S_0\left(1 - \frac{r}{g}\right)}{g-r} = \frac{1}{g-r}\Delta_1^i + \frac{\alpha^i}{g}S_0$$

leading to (3.28). □

Lemma 3.8 *For $r \neq 0$ and $g \neq 0$ with $g = r$, the wealth V_t^i of the ith household group is given by*

$$V_t^i = e^{rt}\left(\left(t - \frac{1}{r}\right)\Delta_1^i + \frac{1}{r}S_0^i\right) + \alpha^i\left(V_0 - \frac{S_0}{r}\right). \tag{3.29}$$

Proof For $r \neq 0$ and $g \neq 0$ with $g = r$, (3.27) becomes

$$x(t) = e^{rt}\left(x_0 + a\int_0^t 1\,ds + b\int_0^t e^{-rs}\,ds\right)$$
$$= e^{rt}\left(x_0 + at + \frac{b}{r}\right) - \frac{b}{r}.$$

Substituting

$$x = V^i, \quad a = \left(\alpha^i - \beta^i\right)C_0, \quad b = \alpha^i\left(S_0 - rV_0\right),$$

this translates to

$$V_t^i = e^{rt}\left(V_0^i + t\left(\alpha^i - \beta^i\right)C_0 + \frac{1}{r}\alpha^i\left(S_0 - rV_0\right)\right) - \frac{1}{r}\alpha^i\left(S_0 - rV_0\right).$$

With

$$\begin{aligned} V_0^i + \frac{1}{r}\alpha^i\left(S_0 - rV_0\right) &= \frac{1}{r}\left(rV_0^i + \alpha^i\left(S_0 - rV_0\right)\right) \\ &= \frac{1}{r}\left(R_0^i + \alpha^i\left(S_0 - R_0\right)\right) \\ &= \frac{1}{r}\left(R_0^i + \alpha^i\left(W_0 - C_0\right)\right) \\ &= \frac{1}{r}\left(R_0^i + W_0^i - C_0^i - \left(\alpha^i - \beta^i\right)C_0\right) \\ &= \frac{1}{r}\left(S_0^i - \Delta_1^i\right), \end{aligned}$$

we are led to (3.29). □

The following proposition collects the results of this section:

Proposition 3.9 *In a circular flow economy with constant parameters, the wealth V_t^i of a household group $i = 1, \ldots, N$ is given by*

$$V_t^i = \begin{cases} tS_0^i + V_0^i & (r = 0,\ g = 0) \\ e^{gt}\frac{1}{g}S_0^i + \left(V_0^i - \frac{1}{g}S_0^i\right) & (r = 0,\ g \neq 0) \\ e^{rt}\left(\frac{1}{r}\Delta_1^i + \Delta_2^i\right) + t\alpha^i S_0 + \left(V_0^i - \left(\frac{1}{r}\Delta_1^i + \Delta_2^i\right)\right) & (r \neq 0,\ g = 0) \\ e^{rt}\left(\left(t - \frac{1}{r}\right)\Delta_1^i + \frac{1}{r}S_0^i\right) + \xi^i & (r \neq 0,\ g \neq 0,\ r = g) \\ e^{rt}\left(\frac{1}{r-g}\Delta_1^i + \Delta_2^i\right) + e^{gt}\left(\frac{1}{g-r}\Delta_1^i + \frac{\alpha^i}{g}S_0\right) + \xi^i & (r \neq 0,\ g \neq 0,\ r \neq g) \end{cases} \tag{3.30}$$

with

$$\xi^i = \alpha^i \left(V_0 - \frac{S_0}{g} \right) \tag{3.31}$$

and

$$\Delta_1^i = \left(\alpha^i - \beta^i\right) C_0, \tag{3.32}$$
$$\Delta_2^i = \left(\upsilon^i - \alpha^i\right) V_0.$$

□

The capital income of the ith household group is

$$R_t^i = r V_t^i,$$

where V_t^i is given by the corresponding expression in Proposition 3.9 depending on the parameters r and g.

Based on (3.15), the savings amount S_t^i of group i at time t is given by the derivative of V_t^i with respect to time. For example, for the case $r \neq 0$ and $g \neq 0$ with $g \neq r$, we have

$$S_t^i = \dot{V}_t^i = e^{rt} \left(\frac{r}{r-g} \Delta_1^i + r \Delta_2^i \right) + e^{gt} \left(\frac{g}{g-r} \Delta_1^i + \alpha^i S_0 \right).$$

Furthermore,

$$C_t^i = \beta^i C_0 e^{gt},$$

using (3.2) and (3.3). With Lemma 3.3 and (3.3), we obtain

$$W_t^i = \begin{cases} -t\alpha^i r S_0 + W_0^i & (g = 0) \\ e^{gt}\alpha^i \left(Y_0 - \frac{r}{g} S_0\right) + \alpha^i \left(\frac{r}{g} S_0 - r V_0\right) & (g \neq 0) \, . \end{cases} \tag{3.33}$$

If $g > 0$ and

$$Y_0 > \frac{r}{g} S_0, \tag{3.34}$$

then W_t^i is increasing since $\alpha^i > 0$ for all $i = 1, \ldots, N$.

3.4 Concentration of Wealth

Under what conditions do the wealth developments of household groups drift further and further apart over time? In this section, we examine this question for a circular flow economy with constant parameters.

Lemma 3.10 *We consider a circular flow economy with constant parameters and assume $g \geq 0$. Then there is at least one household group l with*

$$\lim_{t\to\infty} V_t^l = \infty.$$

Proof According to Lemma 3.1 we have $\lim_{t\to\infty} V_t = \infty$ for $g \geq 0$. □

Lemma 3.11 *For each household group $1 \leq i \leq N$, either*

$$\lim_{t\to\infty} V_t^i = c$$

for some constant c or

$$\lim_{t\to\infty} V_t^i = \pm\infty.$$

Proof This follows from (3.30) using

$$e^{at} \to \begin{cases} 0 & (a < 0) \\ 1 & (a = 0) \\ \infty & (a > 0) \end{cases}$$

for $t \to \infty$. □

Definition 3.12 We say that a **concentration of wealth** occurs in an economy if there are two household groups $1 \leq k,\ l \leq N$ such that

$$V_t^k \leq c$$

for some constant c and

$$\lim_{t\to\infty} V_t^l = +\infty,$$

or if

$$\lim_{t\to\infty} V_t^i = +\infty$$

for all $1 \leq i \leq N$, but

$$\lim_{t\to\infty} \frac{V_t^l}{V_t^k} = +\infty.$$

In particular, $V_t^k \leq c$ if $\lim_{t\to\infty} V_t^k = -\infty$, because then there is a $T > 0$ with $V_t^k < 0$ for all $t > T$, and the continuous function $t \to V_t^k$ restricted to the compact set $[0, T]$ is bounded.

Wealth concentration can therefore be interpreted as the emergence of ever-increasing inequality within the wealth distribution of an economy over time. We now examine conditions under which wealth concentration occurs in a circular flow economy.

Lemma 3.13 *Let arbitrary λ, $\mu \in \mathbb{R}$ be fixed. Then there are indices $1 \leq k, l \leq N$ with*

$$\lambda\Delta_1^k + \mu\Delta_2^k < 0 \quad \textit{and} \quad \lambda\Delta_1^l + \mu\Delta_2^l > 0$$

if and only if not all $\lambda\Delta_1^i + \mu\Delta_2^i$, $1 \leq i \leq N$, vanish.

Proof The assertion follows from $\sum_{i=1}^N \Delta_1^i = \sum_{i=1}^N \Delta_2^i = 0$. □

In particular, there is some index k with $\Delta_1^k < 0$ $\left(\Delta_2^k < 0\right)$ if and only if there is some index l with $\Delta_1^l > 0$ $\left(\Delta_2^l > 0\right)$.

We come to an important result:

Proposition 3.14 *Assume $r \geq g \geq 0$ and $r > 0$ in a circular flow economy with constant parameters.*

1. *Case $r > g \geq 0$. If there are two household groups $1 \leq k, l \leq N$ with*

$$\begin{aligned} \Delta_1^k < 0, \quad \Delta_1^l > 0, \\ \Delta_2^k < 0, \quad \Delta_2^l > 0, \end{aligned} \tag{3.35}$$

then wealth concentration occurs.

2. *Case $r = g > 0$. If there are two household groups $1 \leq k, l \leq N$ with*

$$\Delta_1^k < 0, \quad \Delta_1^l > 0, \tag{3.36}$$

then wealth concentration occurs.

Proof

1. In case $r > g > 0$, (3.30) implies

$$V_t^i = e^{rt}\left(\left(\frac{1}{r-g}\Delta_1^i + \Delta_2^i\right) + e^{-(r-g)t}\left(\frac{1}{g-r}\Delta_1^i + \frac{\alpha^i}{g}S_0\right) + e^{-rt}\xi^i\right)$$

for $1 \leq i \leq N$. According to (3.35),

$$\frac{1}{r-g}\Delta_1^k + \Delta_2^k < 0, \qquad \frac{1}{r-g}\Delta_1^l + \Delta_2^l > 0,$$

so that $\lim_{t\to\infty} V_t^k = -\infty$ and $\lim_{t\to\infty} V_t^l = \infty$. The case $r > g = 0$ is treated similarly.

2. If $r = g > 0$, then (3.30) implies

$$V_t^i = e^{rt}\left(t\Delta_1^i + \frac{1}{r}\left(S_0^i - \Delta_1^i\right)\right) + \xi^i$$

for $1 \leq i \leq N$. Thus, (3.36) leads to $\lim_{t\to\infty} V_t^k = -\infty$ and $\lim_{t\to\infty} V_t^l = \infty$.

□

Proposition 3.14 is in line with the statement by T. Piketty in [17], based on empirical studies, according to which increasing inequality within economies is observed if $r > g$.

In the previous proposition, the signs of Δ_1^i and Δ_2^i proved to be critical for the occurrence of wealth concentration. We now interpret these signs and first note that the inequalities in (3.35) are equivalent to

$$\upsilon^k < \alpha^k < \beta^k$$

$$\upsilon^l > \alpha^l > \beta^l.$$

Definition 3.15 The inequalities

$$\upsilon^k < \alpha^k < \beta^k \tag{3.37}$$

characterize poorer households whose share of wealth is lower than their share of earned income, which in turn is lower than their share of consumer spending. A household group with the characteristics (3.37) is called a **lower household group**.

In contrast, the inequalities

$$\upsilon^l > \alpha^l > \beta^l \tag{3.38}$$

characterize wealthy households. Their shares of wealth are greater than their shares of labor income, and these shares are in turn greater than their shares of consumer spending. A household group with the characteristics (3.38) is called an **upper household group**.

According to Lemma 3.1, the total wealth grows unrestrictedly if $g \geq 0$. In particular, this happens if $r \geq g \geq 0$ and $r > 0$. However, if lower and upper household groups exist in the economy, Proposition 3.14 states that wealth will become concentrated and the wealth gap between household groups in the economy widens over time.

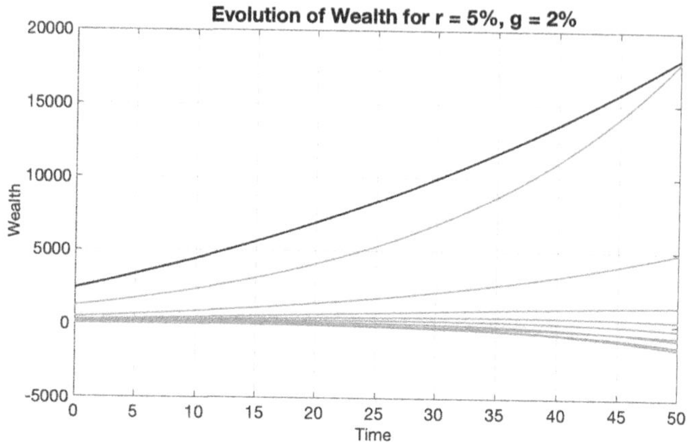

Fig. 3.3 Wealth development in a circular flow economy with ten household groups for $r = 5\%$ and $g = 2\%$. Total wealth of the economy is shown in black and the wealth of the individual household groups in gray

The situation described is illustrated in Fig. 3.3 for the parameter constellation $r = 5\%$ and $g = 2\%$. For a circular flow economy with ten household groups, the wealth development of the individual groups is shown in gray and the development of total wealth in black. We see that the wealth of the individual household groups evolve very differently. Compared to total wealth, the wealth of at least one of the upper household groups is growing more strongly, while the lower groups are increasingly getting into debt.

Even if $g > r > 0$, wealth concentration may occur:

Proposition 3.16 *Assume $g > r > 0$. For each household group k with*

$$\frac{1}{g-r}\Delta_1^k + \frac{\alpha^k}{g} S_0 < 0, \tag{3.39}$$

we have

$$\lim_{t\to\infty} V_t^k = -\infty.$$

Thus, in a circular flow economy with constant parameters, wealth concentration occurs if $g > r > 0$ and if there is at least one household group k for which (3.39) is fulfilled.

Proof If $g > r > 0$, (3.30) implies

$$V_t^i = e^{gt}\left(\left(\frac{1}{g-r}\Delta_1^i + \frac{\alpha^i}{g}S_0\right) + e^{-(g-r)t}\left(\frac{1}{r-g}\Delta_1^i + \Delta_2^i\right) + e^{-gt}\xi^i\right)$$

for $1 \leq i \leq N$. If for some k

$$\lim_{t\to\infty}\left(\left(\frac{1}{g-r}\Delta_1^k+\frac{\alpha^k}{g}S_0\right)+e^{-(g-r)t}\left(\frac{1}{r-g}\Delta_1^k+\Delta_2^k\right)+e^{-gt}\xi^k\right)$$
$$=\frac{1}{g-r}\Delta_1^k+\frac{\alpha^k}{g}S_0<0,$$

then

$$\lim_{t\to\infty} V_t^k=-\infty.$$

On the other hand, according to Lemma 3.10 there is at least one l with $\lim_{t\to\infty} V_t^l=\infty$. □

In the proof above, we might argue alternatively that $\sum_{i=1}^N\left(\frac{1}{g-r}\Delta_1^i+\frac{\alpha^i}{g}S_0\right)=\frac{1}{g}S_0>0$ implies the existence of a group l with $\frac{1}{g-r}\Delta_1^l+\frac{\alpha^l}{g}S_0>0$, so that $\lim_{t\to\infty} V_t^l=\infty$.

The condition (3.39) in Proposition 3.16 can be slightly weakened:

Corollary 3.17 *Assume that $g>r>0$ and that there is a household group k with*

$$\frac{1}{g-r}\Delta_1^k+\frac{\alpha^k}{g}S_0\leq 0. \tag{3.40}$$

Then a concentration of wealth occurs in the economy.

Proof If $\frac{1}{g-r}\Delta_1^k+\frac{\alpha^k}{g}S_0<0$, the assertion follows from Proposition 3.16.

For the remainder of the proof choose l with $\frac{1}{g-r}\Delta_1^l+\frac{\alpha^l}{g}S_0>0$, so that $\lim_{t\to\infty} V_t^l=\infty$. Let us now assume that

$$\frac{1}{g-r}\Delta_1^k+\frac{\alpha^k}{g}S_0=0$$

for some k, so that

$$V_t^k=e^{rt}\left(\frac{1}{r-g}\Delta_1^k+\Delta_2^k\right)+\xi^k.$$

If

$$\frac{1}{r-g}\Delta_1^k+\Delta_2^k<0,$$

we conclude $\lim_{t\to\infty} V_t^k=-\infty$.

If

$$\frac{1}{r-g}\Delta_1^k + \Delta_2^k = 0,$$

then $V_t^k = \xi^k$ and V_t^k is bounded.

Finally, if

$$\frac{1}{r-g}\Delta_1^k + \Delta_2^k > 0,$$

then $\lim_{t\to\infty} V_t^k = +\infty$. But

$$\lim_{t\to\infty} e^{(g-r)t}\left(\frac{1}{g-r}\Delta_1^l + \frac{\alpha^l}{g}S_0\right) = +\infty$$

implies

$$\lim_{t\to\infty}\frac{V_t^l}{V_t^k} = \frac{\left(\frac{1}{r-g}\Delta_1^l + \Delta_2^l\right) + e^{(g-r)t}\left(\frac{1}{g-r}\Delta_1^l + \frac{\alpha^l}{g}S_0\right) + e^{-rt}\xi^l}{\left(\frac{1}{r-g}\Delta_1^k + \Delta_2^k\right) + e^{-rt}\xi^k} = +\infty.$$

□

Corollary 3.18 *Assume $g > r > 0$ in a circular flow economy with constant parameters. If $\Delta_1^k < 0$ holds for a household group k, then wealth concentration occurs for g sufficiently close to r.*

Proof If $\Delta_1^k < 0$ for some k, then

$$\lim_{g\downarrow r}\frac{1}{g-r}\Delta_1^k = -\infty.$$

Therefore, (3.39) is satisfied for $g > r > 0$ and for g sufficiently close to r. Thus, the assertion follows from Corollary 3.17. □

So far, we have discussed conditions under which wealth concentration occurs in a circular flow economy. We are now investigating the circumstances under which this can be avoided.

Proposition 3.19 *In a circular flow economy with constant parameters, a concentration of wealth will be avoided if the following applies:*

1. $r = 0: \; S_0^i > 0$ *for all* i.
2. $r > g = 0: \; \Delta_1^i + r\Delta_2^i = 0$ *for all* i.
3. $r > g > 0: \; \frac{1}{r-g}\Delta_1^i + \Delta_2^i = 0$ *and* $\frac{1}{g-r}\Delta_1^i + \frac{\alpha^i}{g}S_0 > 0$ *for all* i.
4. $r = g > 0: \; \Delta_1^i = 0$ *and* $S_0^i > 0$ *for all* i.
5. $g > r > 0: \; \frac{1}{g-r}\Delta_1^i + \frac{\alpha^i}{g}S_0 > 0$ *for all* i.

Proof All statements follow directly from (3.30). □

Conditions 2, 3, and 4 in Proposition 3.19 require special parameter constellations in the economy that do not seem suitable for describing realistic situations. Therefore, we concentrate on conditions 1 and 5, which we interpret as follows:

- In the case $r = 0$, there is no concentration of wealth if each household group can build up savings.
- In the case $g > r > 0$, there is no concentration of wealth if the inequality

$$\frac{1}{g-r}\Delta_1^i + \frac{\alpha^i}{g}S_0 > 0 \tag{3.41}$$

is fulfilled for each household group $1 \leq i \leq N$. Under these circumstances

$$\lim_{t\to\infty} V_t^i = \infty$$

for each household group i and

$$\lim_{t\to\infty} \frac{V_t^i}{V_t^j} = \frac{\frac{1}{g-r}\Delta_1^i + \frac{\alpha^i}{g}S_0}{\frac{1}{g-r}\Delta_1^j + \frac{\alpha^j}{g}S_0}$$

for every two household groups i and j.

We will now examine when the inequality (3.41) is fulfilled. First of all,

$$\begin{aligned}
\Delta_1^i + \alpha^i S_0 &= \alpha^i\,(C_0 + S_0) - \beta^i C_0 \\
&= \alpha^i Y_0 - C_0^i \\
&= \alpha^i\,(W_0 + R_0) - C_0^i \\
&= W_0^i - C_0^i + r\alpha^i V_0.
\end{aligned}$$

If $g \neq r$, this implies

$$\frac{1}{g-r}\Delta_1^i + \frac{\alpha^i}{g}S_0 = \left(\frac{1}{g}\Delta_1^i + \frac{\alpha^i}{g}S_0\right) + \left(\frac{1}{g-r} - \frac{1}{g}\right)\Delta_1^i$$

$$= \frac{1}{g}\left(W_0^i - C_0^i + r\alpha^i V_0 + \frac{r}{g-r}\Delta_1^i\right).$$

If $g > r > 0$ and $\Delta_1^i \geq 0$, then

$$\frac{1}{g-r}\Delta_1^i + \frac{\alpha^i}{g}S_0 > \frac{1}{g}\left(W_0^i - C_0^i\right).$$

Therefore, (3.41) is fulfilled if household group i is able to finance its consumption expenditure C_0^i with its labor income W_0^i, i.e., if $W_0^i \geq C_0^i$.

If $\Delta_1^i < 0$ and $W_0^i \geq C_0^i$, we obtain for r fixed and $g > r$ sufficiently large

$$\frac{1}{g-r}\Delta_1^i + \frac{\alpha^i}{g}S_0 \geq \frac{r}{g}\left(\alpha^i V_0 + \frac{1}{g-r}\Delta_1^i\right) > 0.$$

In a circular flow economy with constant parameters, wealth concentration can therefore be avoided if:

- Either no capital income occurs, $r = 0$, and each household group can save
- Or if perpetual growth occurs with a growth rate g greater than the capital return rate $r > 0$. In addition, (3.41) must be fulfilled for each household group. For this, it is sufficient that $W_0^i \geq C_0^i$ for each i and that in the case of $\Delta_1^i < 0$ economic growth g is sufficiently large compared to r.

At the end of Sect. 2.1, it was qualitatively derived from the relationship $Y = W + R$ that an economy is forced to grow if capital income is permitted without restriction. This statement is quantified here in the context of circular flow models with constant parameters.

The situation in which there is no concentration of wealth is illustrated in Fig. 3.4. Here again, the wealth of the household groups is shown together with total wealth. The initial data correspond to those on which Fig. 3.3 is based. For Fig. 3.4, however, $r = 2\%$ and $g = 5\%$ were assumed. In addition to that, the initial data were chosen so that (3.41) is fulfilled for each household group, and we see that the wealth of each household group increases over time.

3.5 Interest Transfer

Interest transfer Z_t^i of a household group i was defined in (2.24) as the difference between interest income R_t^i of that group and its interest payments Π_t^i that are included in its consumption expenditures. The latter quantity was modeled as $\Pi_t^i = \beta^i R_t$, so that

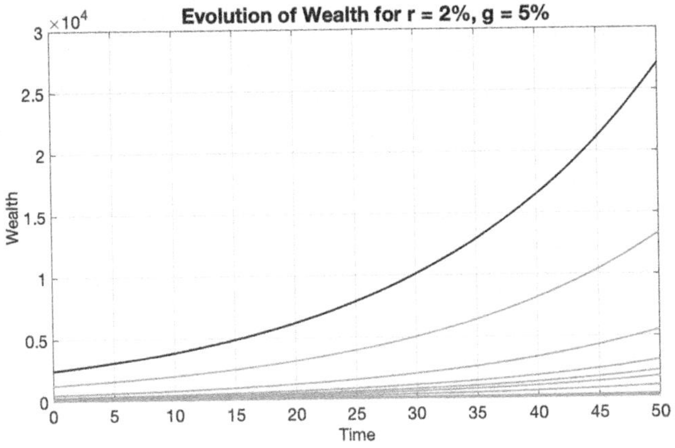

Fig. 3.4 Wealth development in a circular flow economy with ten household groups for $r = 2\%$ and $g = 5\%$. Total wealth of the economy is shown in black and the wealth of the individual household groups in gray

$$Z_t^i = R_t^i - \Pi_t^i = r\left(V_t^i - \beta^i V_t\right). \tag{3.42}$$

Proposition 3.20 *In a circular flow economy with constant parameters, the interest transfer Z_t^i of a household group $i = 1, \ldots, N$ is given by*

$$Z_t^i = \begin{cases} 0 & (r = 0,\ g = 0) \\ 0 & (r = 0,\ g \neq 0) \\ e^{rt}\left(\Delta_1^i + r\Delta_2^i\right) + tr\frac{S_0}{C_0}\Delta_1^i + \left(R_0^i - \beta^i R_0 - \left(\Delta_1^i + r\Delta_2^i\right)\right) & (r \neq 0,\ g = 0) \\ e^{rt}\left((tr - 1)\Delta_1^i + S_0^i - \beta^i S_0\right) + \frac{1}{C_0}(R_0 - S_0)\Delta_1^i & (r \neq 0,\ g \neq 0, \\ & \quad r = g) \\ e^{rt}\left(\frac{r}{r-g}\Delta_1^i + r\Delta_2^i\right) + e^{gt}\left(\frac{r}{g-r} + \frac{r}{g}\frac{S_0}{C_0}\right)\Delta_1^i + \frac{1}{C_0}\left(R_0 - \frac{r}{g}S_0\right)\Delta_1^i & (r \neq 0,\ g \neq 0, \\ & \quad r \neq g)\,. \end{cases} \tag{3.43}$$

Proof We will prove the result for $r \neq 0$, $g \neq 0$, and $r \neq g$, and the other cases are treated similarly. By Lemma 3.1 and Proposition 3.9 we have

$$V_t = e^{gt}\frac{1}{g}S_0 + \left(V_0 - \frac{S_0}{g}\right)$$

and

$$V_t^i = e^{rt}\left(\frac{1}{r-g}\Delta_1^i + \Delta_2^i\right) + e^{gt}\left(\frac{1}{g-r}\Delta_1^i + \frac{\alpha^i}{g}S_0\right) + \alpha^i\left(V_0 - \frac{S_0}{g}\right).$$

We thus find

$$Z_t^i = r\left(V_t^i - \beta^i V_t\right)$$
$$= e^{rt}\left(\frac{r}{r-g}\Delta_1^i + r\Delta_2^i\right) + e^{gt}\left(\frac{r}{g-r} + \frac{r}{g}\frac{S_0}{C_0}\right)\Delta_1^i + \frac{r}{C_0}\left(V_0 - \frac{S_0}{g}\right)\Delta_1^i.$$

□

From definition (3.42) we conclude $\sum_{i=1}^N Z_t^i = r\left(\sum_{i=1}^N V_t^i - V_t\sum_{i=1}^N \beta^i\right) = 0$, and this also follows from each formula in (3.43) because of $\sum_{i=1}^N \Delta_1^i = \sum_{i=1}^N \Delta_2^i = 0$.

If $g > r > 0$, we write Z_t^i in (3.43) as

$$Z_t^i = e^{gt}\left[\left(\frac{r}{g-r} + \frac{r}{g}\frac{S_0}{C_0}\right)\Delta_1^i + e^{(r-g)t}\left(\frac{r}{r-g}\Delta_1^i + r\Delta_2^i\right)\right.$$
$$\left. + e^{-gt}\frac{r}{C_0}\left(V_0 - \frac{S_0}{g}\right)\Delta_1^i\right].$$

For the given choice of parameters, we have $\frac{r}{g-r} + \frac{r}{g}\frac{S_0}{C_0} > 0$. If we now assume that k is a lower and l an upper household group, then we conclude with (3.35)

$$Z_t^k < 0, \quad Z_t^l > 0$$

for t sufficiently large. Furthermore,

$$\lim_{t\to\infty} Z_t^k = -\infty, \qquad \lim_{t\to\infty} Z_t^l = +\infty.$$

If $r > g > 0$, we write Z_t^i in (3.43) as

$$Z_t^i = e^{rt}\left[\left(\frac{r}{r-g}\Delta_1^i + r\Delta_2^i\right) + e^{(g-r)t}\left(\frac{r}{g-r} + \frac{r}{g}\frac{S_0}{C_0}\right)\Delta_1^i\right.$$
$$\left. + e^{-rt}\frac{r}{C_0}\left(V_0 - \frac{S_0}{g}\right)\Delta_1^i\right]$$

and find analogously

$$Z_t^k < 0, \quad Z_t^l > 0$$

for t sufficiently large and

$$\lim_{t\to\infty} Z_t^k = -\infty, \qquad \lim_{t\to\infty} Z_t^l = +\infty.$$

If $r = g > 0$, we have

$$Z_t^i = e^{rt}\left[\left((tr-1)\,\Delta_1^i + S_0^i - \beta^i S_0\right) + e^{-rt}\frac{1}{C_0}\left(R_0 - S_0\right)\Delta_1^i\right],$$

and if $r > g = 0$, we have

$$Z_t^i = e^{rt}\left[\left(\Delta_1^i + r\Delta_2^i\right) + e^{-rt}tr\frac{S_0}{C_0}\Delta_1^i + e^{-rt}\left(R_0^i - \beta^i R_0 - \left(\Delta_1^i + r\Delta_2^i\right)\right)\right].$$

In both cases,

$$Z_t^k < 0, \quad Z_t^l > 0$$

for sufficiently large t and

$$\lim_{t\to\infty} Z_t^k = -\infty, \quad \lim_{t\to\infty} Z_t^l = +\infty.$$

These results show that interest transfer is an effective redistribution mechanism that is always at work regardless of the relation between $r > 0$ and $g \geq 0$. As a result, the lower household groups are forced to make ever higher interest transfers to the upper household groups.

However, if economic growth is large enough, it may well be that these payments can be afforded by the lower household groups without them having to take on debt. To illustrate this, let us consider the case $g > r > 0$, in which the wealth of a household group is—according to (3.30)—given by

$$V_t^i = e^{gt}\left[\left(\frac{1}{g-r}\Delta_1^i + \frac{\alpha^i}{g}S_0\right) + e^{(r-g)t}\left(\frac{1}{r-g}\Delta_1^i + \Delta_2^i\right)\right.$$
$$\left. + e^{-gt}\alpha^i\left(V_0 - \frac{S_0}{g}\right)\right].$$

If $\frac{1}{g-r}\Delta_1^i + \frac{\alpha^i}{g}S_0 > 0$, V_t^i is positive for t sufficiently large. In this case, the household group is able to make the transfer payments without incurring debt.

However, if $r \geq g \geq 0$ and $r > 0$, then Proposition 3.14 shows that wealth concentration occurs in the economy, and in this case the lower household groups become increasingly indebted over time in favor of the upper household groups.

These results provide justification for the ubiquitous demand for eternal growth. If capital gains cannot be restricted in an appropriate manner, then the analysis of the circular flow model with constant parameters indicates that the only way to protect the economy from wealth concentration is to generate eternal growth.

However, perpetual growth that is not generated solely by ever-increasing inflation is not possible in a world with finite resources, and we are confronted with the following unpleasant situation: The unrealistic demand for continuous growth is being misused to maintain the current treatment of capital income. However, if the results of this chapter can be transferred to real economies, at least qualitatively, then allowing unlimited capital gains will lead to economic destabilization in the long run.

In Chap. 8, we will see that the current tax system does not prevent wealth concentration, but possible ways out of this dilemma will be discussed.

Chapter 4
Lorenz Curves and Gini Coefficients

In economics, Lorenz curves are used for the graphical illustration of the distribution of a macroeconomic quantity among the household groups that make up an economy. Originally, this concept was developed by Max O. Lorenz in 1905 for representing the inequality of a wealth distribution. Later in 1912, Corrado Gini published a quantitative measure based on the Lorenz curve to express the amount of inequality numerically. This measure became henceforth to be known as the Gini coefficient.

4.1 Lorenz Curves

Assume that the economy is divided into N groups of equal numbers of households. Assume further an amount Y^i of some economic quantity for each group $i = 1, \ldots, N$ and that $Y = \sum_{i=1}^{N} Y^i > 0$. Let

$$y_i = \frac{\sum_{k=1}^{i} Y^k}{Y}$$

be the proportion of this quantity belonging to the first i groups. By definition, $y_0 = 0$ and $y_N = 1$.

With $x_i = \frac{i}{N}$ the polygon curve defined by the points

$$(x_i, y_i) \quad (i = 0, \ldots, N)$$

is called the **Lorenz curve** for the quantity under consideration, and it connects the point $(0, 0)$ with the point $(1, 1)$ in the plane. A Lorenz curve may be described using a function $L : [0, 1] \to \mathbb{R}$, so that the points on the Lorenz curve are given by $(x, L(x))$ for $x \in [0, 1]$ with $L(x_i) = y_i$ for $i = 0, \ldots, N$.

J. Kremer, *Capital Income and Concentration of Wealth*, Springer Studies in Alternative Economics, https://doi.org/10.1007/978-3-658-50715-2_4

If the considered quantity is distributed evenly among the household groups, i.e., if $Y^i = \frac{1}{N}Y$ for $i = 1, \ldots, N$, then

$$y_i = \frac{i}{N},$$

so in this case the points (x_i, y_i) are located on the diagonal of the unit square.

If, on the other hand, $Y^1 < \cdots < Y^N$, so that the first groups have a smaller amount of the quantity under consideration than the last groups, the Lorenz curve runs below the diagonal, and both curves enclose a positive area. This is explained in Sect. 4.2 below.

4.2 The Gini Coefficient

The **Gini coefficient** of an economic quantity is defined as twice the area enclosed by the corresponding Lorenz curve and the diagonal in the unit square. In the case $0 \le Y^1 \le \cdots \le Y^N$ and $Y > 0$, this definition results in numbers between 0 and 1, see Lemma 4.2. Gini coefficients near 0 characterize uniform distributions, and the higher the Gini coefficient, the greater the inequality between household groups. If some of the Y^i even happen to be negative, as might be the case for a wealth distribution, then Gini coefficients larger than 1 may occur.

Consider a function $L : [0, 1] \to \mathbb{R}$ so that the Lorenz curve $x \to (x, L(x))$ connects the $N + 1$ points (x_i, y_i) for $i = 0, \ldots, N$ by straight line sections. Furthermore, we assume $(x_0, y_0) = (0, 0)$ and $(x_N, y_N) = (1, 1)$. The area enclosed by the diagonal and the Lorenz curve is given by

$$\int_0^1 |x - L(x)| \, dx.$$

For the important special case that the Lorenz curve is located below the diagonal, i.e., $y_i \le x_i = \frac{i}{N}$, the above integral simplifies to

$$\int_0^1 (x - L(x)) \, dx = \frac{1}{2} - \int_0^1 L(x) \, dx.$$

The integral of L from 0 to 1 can be written as the sum of the areas of N trapezoids

$$\int_0^1 L(x) \, dx = \sum_{i=1}^{N} \frac{1}{N} \left(\frac{y_{i-1} + y_i}{2} \right)$$
$$= \frac{1}{2N} (y_0 + 2y_1 + \cdots + 2y_{N-1} + y_N)$$

$$= \frac{1}{2N}\left(2y_1 + \cdots + 2y_{N-1} + 2y_N - y_N\right)$$

$$= \frac{1}{N}\sum_{i=1}^{N} y_i - \frac{1}{2N},$$

where $y_0 = 0$ and $y_N = 1$ was used. The Gini coefficient G of a Lorenz curve L is defined as twice the area enclosed between the diagonal and the graph of L. If we assume $L(x) \leq x$, we obtain

$$G = 2\int_0^1 (x - L(x))\,\mathrm{d}x \tag{4.1}$$

$$= 1 - 2\int_0^1 L(x)\,\mathrm{d}x$$

$$= \frac{N+1}{N} - \frac{2}{N}\sum_{i=1}^{N} y_i.$$

Using the partial sums

$$\hat{Y}^i = \sum_{k=1}^{i} Y^k$$

of the Y^i, we write (4.1) as

$$G = \frac{N+1}{N} - \frac{2}{N}\sum_{i=1}^{N}\frac{\hat{Y}^i}{Y} \tag{4.2}$$

$$= \frac{N+1}{N} - \frac{2}{N}\frac{1}{Y}\left(NY^1 + (N-1)Y^2 + \cdots + Y^N\right).$$

The Gini coefficient G can thus be expressed using the values $Y^1, \ldots, Y^N$ of the economic variable under consideration for each household group, and we also call G the **Gini coefficient of the distribution** $Y^1, \ldots, Y^N$.

Definition 4.1 Let $Y_t^1 \ldots, Y_t^N$ be the distribution of an economic variable at time t. Then we define

$$G_t = \frac{N+1}{N} - \frac{2}{N}\frac{1}{Y_t}\left(NY_t^1 + (N-1)Y_t^2 + \cdots + Y_t^N\right). \tag{4.3}$$

If the Lorenz curve associated with $Y_t^1, \ldots, Y_t^N$ is located along or below the diagonal, then G_t coincides with the Gini coefficient of the distribution at time t.

We now specify a condition, so that the Lorenz curve corresponding to a distribution $Y^1, \ldots, Y^N$ runs on or below the diagonal in the unit square.

Lemma 4.2 *If $Y^1 \leq \cdots \leq Y^N$ with $Y > 0$, then*

$$Y^i \leq \frac{1}{N} Y \quad (1 \leq i \leq N),$$

so that

$$y_i = \frac{\hat{Y}^i}{Y} \leq \frac{i}{N} = x_i \quad (1 \leq i \leq N).$$

Thus, the Lorenz curve belonging to the Y^i runs along or below the diagonal in the unit square and

$$G = \frac{N+1}{N} - \frac{2}{N} \frac{1}{Y} \left(N Y^1 + (N-1) Y^2 + \cdots + Y^N \right) \geq 0.$$

Proof Due to the monotonicity of the Y^i, the assumption

$$\hat{Y}^1 = Y^1 > \frac{1}{N} Y$$

leads to $\frac{1}{N} Y < Y^1 \leq \cdots \leq Y^N$ and therefore to

$$\begin{aligned} Y &= \frac{1}{N} Y + (N-1) \frac{1}{N} Y \\ &< Y^1 + (N-1) Y^1 \\ &\leq Y^1 + \left(Y^2 + \cdots + Y^N \right), \end{aligned}$$

contradicting $Y = \sum_{i=1}^{N} Y^i$. Thus, we have $Y^1 \leq \frac{1}{N} Y$.

If we further assume

$$\hat{Y}^2 = Y^1 + Y^2 > \frac{2}{N} Y,$$

then $Y^2 > \frac{1}{N} Y$ because we already know $Y^1 \leq \frac{1}{N} Y$. Consequently, $\frac{1}{N} Y < Y^2 \leq Y^3 \leq \cdots \leq Y^N$ and therefore

$$\begin{aligned} Y &= \frac{2}{N} Y + (N-2) \frac{1}{N} Y \\ &< \left(Y^1 + Y^2 \right) + (N-2) Y^2 \end{aligned}$$

$$\leq \left(Y^1 + Y^2\right) + \left(Y^3 + \cdots + Y^N\right),$$

again contradicting $Y = \sum_{i=1}^{N} Y^i$. Thus, we have $Y^2 \leq \frac{1}{N}Y$. Iteration of this argument leads to $Y^i \leq \frac{1}{N}Y$ for all $1 \leq i \leq N$.

We have shown that the Lorenz curve lies on or below the diagonal. Due to the monotonicity of the integral, the corresponding Gini coefficient is greater than or equal to zero, $G \geq 0$. □

Corollary 4.3 *Assume an index $1 \leq k < N$ with $Y^1 \leq 0, \ldots, Y^k \leq 0$ and $0 \leq Y^{k+1} \leq \cdots \leq Y^N$. Furthermore, suppose $Y > 0$. Then,*

$$Y^i \leq \frac{1}{N}Y \quad (1 \leq i \leq N),$$

so that

$$y_i = \frac{\hat{Y}^i}{Y} \leq \frac{i}{N} = x_i \quad (1 \leq i \leq N).$$

Therefore, the Lorenz curve belonging to the Y^i runs along or below the diagonal and

$$G = \frac{N+1}{N} - \frac{2}{N}\frac{1}{Y}\left(NY^1 + (N-1)\,Y^2 + \cdots + Y^N\right) \geq 0.$$

Proof For each permutation $\pi : \{1, \ldots, k\} \to \{1, \ldots, k\}$, the Lorenz curve L_π of the distribution $Y^{\pi(1)}, \ldots, Y^{\pi(k)}, Y^{k+1}, \ldots, Y^N$ has function values $L_\pi(x_i) \leq 0$ for $i = 1, \ldots, k$. In particular, we obtain $L_\pi(x_i) \leq x_i$ for $i = 1, \ldots, k$. Furthermore, the values $L_\pi(x_k)$ of the Lorenz curves L_π coincide at the point x_k for all permutations π. If π_0 is a permutation with $Y^{\pi_0(1)} \leq \cdots \leq Y^{\pi_0(k)}$, then the prerequisites of Lemma 4.2 are fulfilled for L_{π_0}, and thus the assertion follows from Lemma 4.2. □

Using the next two results, the positivity of (4.3) in Definition 4.1 can be established for a distribution without having to make assumptions about the monotonicity of the members of the distribution.

Lemma 4.4 *Let N numbers $a_1, \ldots, a_N$ be given, so that*

$$\sum_{i=1}^{N} a_i = 0.$$

Further assume an index $1 \leq k < N$ with $a_1 < 0, \ldots, a_k < 0$ and $a_{k+1} \geq 0, \ldots, a_N \geq 0$. In particular, we assume that not all a_i vanish. Then

$$Na_1 + (N-1)\,a_2 + \cdots + a_N < 0.$$

Proof With $a_{k+1} + \cdots + a_N = -(a_1 + \cdots + a_k)$ we calculate

$$\begin{aligned} Na_1 + (N-1)\,a_2 + \cdots + a_N &= Na_1 + (N-1)\,a_2 + \cdots + (N-(k-1))\,a_k \\ &\quad + (N-k)\,a_{k+1} + (N-(k+1))\,a_{k+2} \\ &\quad + \cdots + a_N \\ &< (N-k+1)\,(a_1 + \cdots + a_k) + (N-k) \\ &\quad \times (a_{k+1} + \cdots + a_N) \\ &= (a_1 + \cdots + a_k)\,((N-k+1) - (N-k)) \\ &= a_1 + \cdots + a_k \\ &< 0. \end{aligned}$$

□

Corollary 4.5 *Let N numbers $b_1, \ldots, b_N$ be given, so that*

$$\sum_{i=1}^{N} b_i = 1.$$

Assume that there is an index $1 \leq k < N$ with $b_1 < \frac{1}{N}, \ldots, b_k < \frac{1}{N}$ and $b_{k+1} \geq \frac{1}{N}, \ldots, b_N \geq \frac{1}{N}$. Then

$$\frac{N+1}{N} - \frac{2}{N}\,(Nb_1 + (N-1)\,b_2 + \cdots + b_N) > 0. \tag{4.4}$$

Proof With

$$a_i = b_i - \frac{1}{N}$$

we have

$$\sum_{i=1}^{N} a_i = 0,$$

as well as $a_1 < 0, \ldots, a_k < 0$ and $a_{k+1} \geq 0, \ldots, a_N \geq 0$. Furthermore,

$$\begin{aligned} Na_1 + (N-1)\,a_2 + \cdots + a_N &= Nb_1 + (N-1)\,b_2 + \cdots + b_N \\ &\quad - \frac{1}{N}\,(N + (N-1) + \cdots + 1) \end{aligned}$$

$$= Nb_1 + (N-1)\,b_2 + \cdots + b_N - \frac{N+1}{2}.$$

Thus, with Lemma 4.4 we obtain

$$\frac{N+1}{N} - \frac{2}{N}\,(Nb_1 + (N-1)\,b_2 + \cdots + b_N)$$
$$= -\frac{2}{N}\,(Na_1 + (N-1)\,a_2 + \cdots + a_N) > 0.$$

□

Corollary 4.6 *Let N numbers $b_1 \leq \cdots \leq b_N$ be given with $b_1 < b_N$ and*

$$\sum_{i=1}^{N} b_i = 1.$$

Then

$$\frac{N+1}{N} - \frac{2}{N}\,(Nb_1 + (N-1)\,b_2 + \cdots + b_N) > 0.$$

Proof First of all, $b_1 < 1/N$ follows, because otherwise $b_1 \geq 1/N$ and therefore $b_1 + \cdots + b_N > Nb_1 \geq 1$. Correspondingly, $b_N > 1/N$ also follows, because otherwise $b_N \leq 1/N$ and thus $b_1 + \cdots + b_N < Nb_N \leq 1$. Therefore, the prerequisites of Corollary 4.5 are fulfilled and (4.4) follows. □

4.3 Asymptotic Development of the Gini Coefficients for Wealth Distributions

For a circular flow model with constant parameters, the asymptotic Gini coefficients of the wealth distributions will be examined using G_t in (4.3) for $t \to \infty$.

Prerequisites For the remainder of this section we assume that there are indices $1 \leq k,\, k' < N$ with

$$\Delta_1^1 < 0, \ldots, \Delta_1^k < 0, \quad \Delta_1^{k+1} \geq 0, \ldots, \Delta_1^N \geq 0, \tag{4.5}$$
$$\Delta_2^1 < 0, \ldots, \Delta_2^{k'} < 0, \quad \Delta_2^{k'+1} \geq 0, \ldots, \Delta_2^N \geq 0.$$

In particular, lower and upper household groups exist in the circular flow economies under consideration.

The investigations are carried out within the framework of the following case distinctions:

$$\begin{array}{lll} r=0 & g=0 & \\ r=0 & g>0 & \\ r>0 & g=0 & \\ r>0 & g>0 & r=g \\ r>0 & g>0 & r>g \\ r>0 & g>0 & g>r. \end{array}$$

We use (3.5) for total wealth and (3.30) for the wealth of the individual household groups.

Case $r = g = 0$ Here, we have

$$\frac{V_t^i}{V_t} = \frac{tS_0^i + V_0^i}{tS_0 + V_0} = \frac{S_0^i + V_0^i/t}{S_0 + V_0/t}.$$

For $t \to \infty$ we obtain

$$\frac{V_t^i}{V_t} \to \frac{S_0^i}{S_0}$$

so that

$$\lim_{t\to\infty} G_t = \frac{N+1}{N} - \frac{2}{N}\frac{1}{S_0}\left(NS_0^1 + (N-1)\,S_0^2 + \cdots + S_0^N\right).$$

Assuming an index $1 \le k < N$ with

$$S_0^1 < \frac{1}{N}S_0, \ldots, S_0^k < \frac{1}{N}S_0 \quad \text{and} \quad S_0^{k+1} \ge \frac{1}{N}S^0, \ldots, S_0^N \ge \frac{1}{N}S_0,$$

or assuming $S_0^1 \le \cdots \le S_0^N$ with $S_0^1 < S_0^N$, then we conclude from Corollary 4.5 or from Corollary 4.6

$$\frac{N+1}{N} - \frac{2}{N}\frac{1}{S_0}\left(NS_0^1 + (N-1)\,S_0^2 + \cdots + S_0^N\right) > 0.$$

Case $g > r = 0$ In this case, we have

$$\frac{V_t^i}{V_t} = \frac{e^{gt}\frac{1}{g}S_0^i + \left(V_0^i - \frac{S_0^i}{g}\right)}{e^{gt}\frac{1}{g}S_0 + \left(V_0 - \frac{S_0}{g}\right)}$$

$$= \frac{S_0^i + e^{-gt}\left(gV_0^i - S_0^i\right)}{S_0 + e^{-gt}\left(gV_0 - S_0\right)}.$$

Thus, for $t \to \infty$

$$\frac{V_t^i}{V_t} \to \frac{S_0^i}{S_0},$$

and we get the same asymptotic result as in the previous case:

$$\lim_{t\to\infty} G_t = \frac{N+1}{N} - \frac{2}{N}\frac{1}{S_0}\left(NS_0^1 + (N-1)\,S_0^2 + \cdots + S_0^N\right).$$

Case $r > g = 0$ Using $\Delta^i = \Delta_1^i + r\Delta_2^i$, the wealth V_t^i of a household group i at time t is given by

$$V_t^i = e^{rt}\frac{1}{r}\Delta^i + t\alpha^i S_0 + \left(V_0^i - \frac{1}{r}\Delta^i\right),$$

and therefore

$$\frac{V_t^i}{V_t} = e^{rt}\frac{1}{tr}\Delta^i\frac{1}{S_0 + V_0/t} + \frac{\alpha^i S_0 + \left(V_0^i - \frac{1}{r}\Delta^i\right)/t}{S_0 + V_0/t}.$$

Thus, for t sufficiently large

$$\frac{V_t^i}{V_t} \approx e^{rt}\frac{1}{trS_0}\Delta^i + \alpha^i,$$

and G_t is approximately given by

$$G_t \approx \frac{N+1}{N} - \frac{2}{N}e^{rt}\frac{1}{trS_0}\left(N\Delta^1 + \cdots + \Delta^N\right) - \frac{2}{N}\left(N\alpha^1 + \cdots + \alpha^N\right).$$

From (4.5) and Lemma 4.4 we obtain

$$N\Delta^1 + \cdots + \Delta^N = \left(N\Delta_1^1 + \cdots + \Delta_1^N\right) + r\left(N\Delta_2^1 + \cdots + \Delta_2^N\right) < 0,$$

so that

$$\lim_{t\to\infty} G_t = \infty.$$

Case $r = g > 0$ In this case, the wealth V_t^i of a household group i at time t is given by

$$V_t^i = e^{rt}\left(\left(t - \frac{1}{r}\right)\Delta_1^i + \frac{1}{r}S_0^i\right) + \xi^i,$$

so that

$$\begin{aligned}\frac{V_t^i}{V_t} &= \frac{e^{rt}\left(\left(t - \frac{1}{r}\right)\Delta_1^i + \frac{1}{r}S_0^i\right) + \xi^i}{e^{rt}\frac{1}{r}S_0 + \left(V_0 - \frac{S_0}{r}\right)} \\ &= \frac{\left(\left(t - \frac{1}{r}\right)\Delta_1^i + \frac{1}{r}S_0^i\right) + e^{-rt}\xi^i}{\frac{1}{r}S_0 + e^{-rt}\left(V_0 - \frac{S_0}{r}\right)}.\end{aligned}$$

For t sufficiently large

$$\frac{V_t^i}{V_t} \approx \frac{r}{S_0}\left(t - \frac{1}{r}\right)\Delta_1^i + \frac{S_0^i}{S_0},$$

and G_t is approximately given by

$$\begin{aligned}G_t \approx &\frac{N+1}{N} - \frac{2}{N}\frac{r}{S_0}\left(t - \frac{1}{r}\right)\left(N\Delta_1^1 + \cdots + \Delta_1^N\right) \\ &- \frac{2}{N}\frac{1}{S_0}\left(NS_0^1 + (N-1)\,S_0^2 + \cdots + S_0^N\right).\end{aligned}$$

From (4.5) and Lemma 4.4 we conclude

$$N\Delta_1^1 + \cdots + \Delta_1^N < 0,$$

and therefore

$$\lim_{t\to\infty} G_t = \infty.$$

Case $r,\ g > 0,\ r \neq g$ In this case, the wealth V_t^i of a household group i at time t is given by

$$V_t^i = e^{rt}\left(\frac{1}{r-g}\Delta_1^i + \Delta_2^i\right) + e^{gt}\left(\frac{1}{g-r}\Delta_1^i + \frac{\alpha^i}{g}S_0\right) + \xi^i$$

so that

$$\frac{V_t^i}{V_t} = \frac{e^{rt}\left(\frac{1}{r-y}\Delta_1^i + \Delta_2^i\right) + e^{gt}\left(\frac{1}{g-r}\Delta_1^i + \frac{\alpha^i}{g}S_0\right) + \xi^i}{e^{gt}\frac{1}{g}S_0 + \left(V_0 - \frac{S_0}{g}\right)}.$$

If $r > g$, then we have for large t

$$\frac{V_t^i}{V_t} \approx e^{(r-g)t}\frac{g}{S_0}\left(\frac{1}{r-y}\Delta_1^i + \Delta_2^i\right) + \frac{g}{g-r}\frac{1}{S_0}\Delta_1^i + \alpha^i,$$

and G_t is approximately given by

$$G_t \approx \frac{N+1}{N} - \frac{2}{N}\left(Nb_t^1 + \cdots + b_t^N\right)$$

with

$$b_t^i = e^{(r-g)t}\frac{g}{S_0}\left(\frac{1}{r-y}\Delta_1^i + \Delta_2^i\right) + \frac{g}{g-r}\frac{1}{S_0}\Delta_1^i + \alpha^i.$$

From (4.5) and Lemma 4.4 we conclude

$$N\Delta_1^1 + \cdots + \Delta_1^N < 0 \quad \text{and} \quad N\Delta_2^1 + \cdots + \Delta_2^N < 0.$$

This implies

$$\lim_{t\to\infty} G_t = +\infty.$$

If $g > r$, then for large t

$$\frac{V_t^i}{V_t} \approx \frac{g}{g-r}\frac{1}{S_0}\Delta_1^i + \alpha^i,$$

so that

$$G_t \approx \left(\frac{N+1}{N} - \frac{2}{N}\left(N\alpha^1 + \cdots + \alpha^N\right)\right) - \frac{2}{N}\frac{g}{g-r}\frac{1}{S_0}\left(N\Delta_1^1 + \cdots + \Delta_1^N\right).$$

Using $N\Delta_1^1 + \cdots + \Delta_1^N < 0$, we conclude

$$-\frac{2}{N}\frac{g}{g-r}\frac{1}{S_0}\left(N\Delta_1^1 + \cdots + \Delta_1^N\right) > 0.$$

Now, assume the relations $\alpha^1 \leq \cdots \leq \alpha^N$ with $\alpha^1 < \alpha^N$ or alternatively $\alpha^1 < \frac{1}{N}, \ldots, \alpha^k < \frac{1}{N}$ and $\alpha^{k+1} \geq \frac{1}{N}, \ldots, \alpha^N \geq \frac{1}{N}$ for some index $1 \leq k < N$. Then, we conclude from Corollary 4.6

$$\frac{N+1}{N} - \frac{2}{N}\left(N\alpha^1 + \cdots + \alpha^N\right) > 0.$$

Thus, we obtain

$$\lim_{t\to\infty} G_t = \frac{N+1}{N} - \frac{2}{N}\frac{g}{g-r}\frac{1}{S_0}\left(N\Delta_1^1 + \cdots + \Delta_1^N\right) - \frac{2}{N}\left(N\alpha^1 + \cdots + \alpha^N\right) > 0.$$

The results achieved are summarized in the following proposition:

Proposition 4.7 *In a circular economy with constant parameters, the following applies under the assumptions made in the above case distinctions:*

$$\lim_{t\to\infty} G_t = \begin{cases} \frac{N+1}{N} - \frac{2}{N}\frac{1}{S_0}\left(NS_0^1 + \cdots + S_0^N\right) & (r = g = 0) \\ \frac{N+1}{N} - \frac{2}{N}\frac{1}{S_0}\left(NS_0^1 + \cdots + S_0^N\right) & (g > r = 0) \\ \infty & (r > g = 0) \\ \infty & (r > g > 0) \\ \infty & (r = g > 0) \\ \frac{N+1}{N} - \frac{2}{N}\left(N\left(\frac{g}{g-r}\frac{1}{S_0}\Delta_1^1 + \alpha^1\right)\right. & \\ \quad \left. + \cdots + \left(\frac{g}{g-r}\frac{1}{S_0}\Delta_1^N + \alpha^N\right)\right) & (g > r > 0). \end{cases}$$

□

Two examples may illustrate Proposition 4.7:

Example 4.8 ($r = 2\%$, $g = 5\%$) In this scenario, economic growth $g = 5\%$ is significantly higher than the rate of return on capital $r = 2\%$. According to Proposition 4.7, the Gini coefficient converges to a finite limit in this case. With the initial data for wealth, labor income, and consumption expenditure specified in the appendix, we find

$$\lim_{t\to\infty} G_t = \frac{N+1}{N} - \frac{2}{N}\left(N\left(\frac{g}{g-r}\frac{1}{S_0}\Delta_1^1 + \alpha^1\right)\right.$$
$$\left. + \cdots + \left(\frac{g}{g-r}\frac{1}{S_0}\Delta_1^{10} + \alpha^{10}\right)\right) = 0,66.$$

With the given initial data we further have

$$\frac{g}{g-r}\frac{1}{S_0}\Delta_1^i + \alpha^i > 0$$

for each household group i. According to Proposition 3.19, wealth concentration does not occur. In Fig. 4.1, Lorenz curves and corresponding Gini coefficients are specified for $t = 0$, $t = 25$, and $t = 50$. △

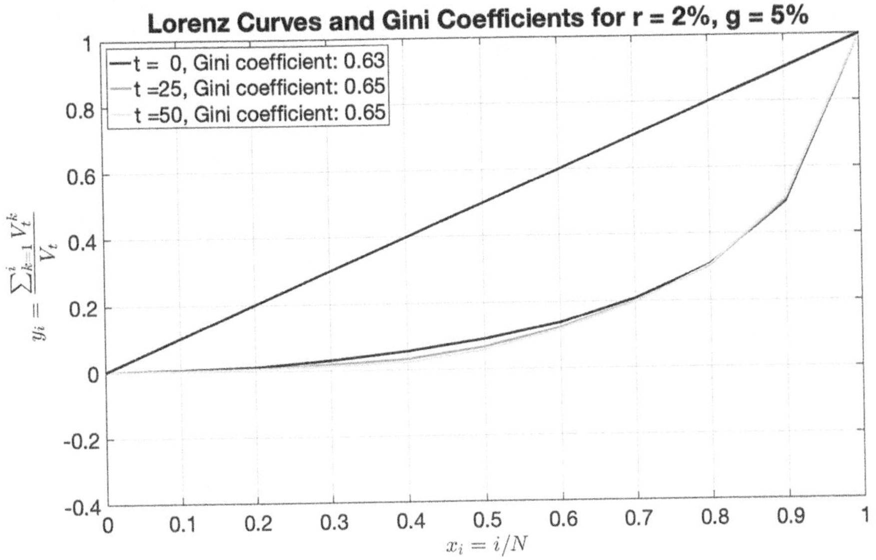

Fig. 4.1 Lorenz curves and Gini coefficients for $r < g$

Example 4.9 (**$r = 5\%$, $g = 2\%$**) The initial data for wealth, labor income, and consumption expenditures of the household groups on which this scenario is based correspond to those of the previous Example 4.8, so that the Lorenz curves in Examples 4.8 and 4.9 coincide at time 0.

In this example, the rate of return on capital $r = 5\%$ is higher than economic growth $g = 2\%$, so that in this case a concentration of wealth occurs in the economy according to Proposition 3.14. Furthermore, with the given initial data, (4.5) is fulfilled, and according to Proposition 4.7 we obtain

$$\lim_{t\to\infty} G_t = \infty.$$

In Fig. 4.2 Lorenz curves and corresponding Gini coefficients are specified for $t = 0$, $t = 25$, and $t = 50$, and we see that a Gini coefficient greater than 1 already occurs for $t = 50$. △

Inequality always occurs in real economies, regardless of the relationship between the rate of return on capital r and economic growth g. According to DIW Weekly Report 40 from 2019 [7], the poorer half of the German population had a share of 1.3% of total wealth in that year. DIW Weekly Report 29 from 2020 [19] states a Gini coefficient of 0.83 for the wealth distribution in Germany.

Example 4.10 Here we consider an economy with $N = 10$ household groups and assume that the first five groups do not possess any wealth. The Lorenz curve for the wealth distribution is assumed to be zero from $x_0 = 0$ to $x_5 = 0.5$ and to increase linearly from 0 to 1 from x_5 to $x_{10} = 1$. In this case, the Gini coefficient is given by

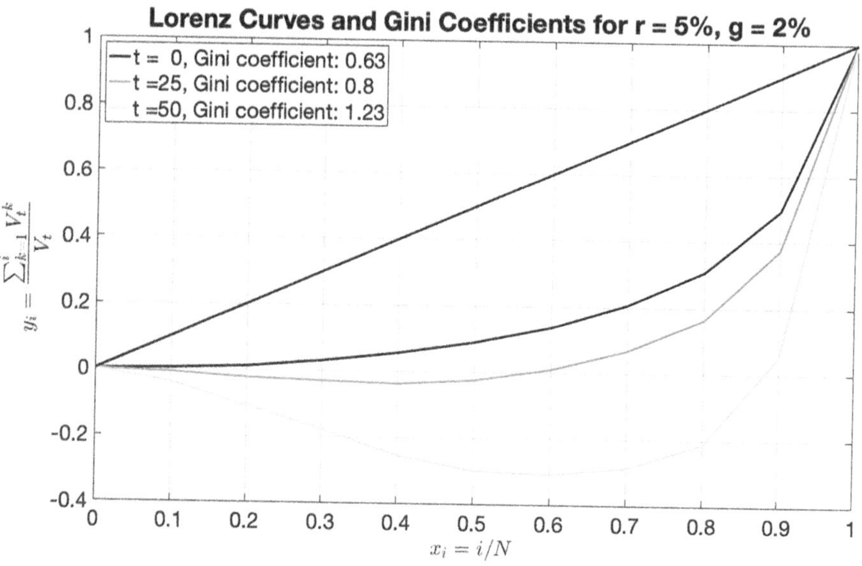

Fig. 4.2 Lorenz curves and Gini coefficients for $r > g$

$$G = 2\left(\frac{1}{2} - \frac{1}{4}\right) = 0.5.$$

Thus, although inequality is very pronounced in this example economy, the Gini coefficient is 0.5, which is not unusually high compared to real values. △

Section 3.4 and Proposition 4.7 show that if $\lim_{t\to\infty} G_t = \infty$ applies, a concentration of wealth occurs in the economy under consideration, i.e., an ever-increasing inequality of the wealth distribution over time.

However, the converse of this statement is not necessarily true. To see this, assume $g > r > 0$ and

$$\frac{1}{g-r}\Delta_1^i + \frac{\alpha^i}{g} S_0 \leq 0$$

for at least one household group i. Then, a concentration of wealth occurs according to Proposition 3.16. Proposition 4.7 shows, however, that G_t remains bounded in this case. So if the Gini coefficient converges to a finite limit for $t \to \infty$, it cannot be concluded that no wealth concentration occurs in the economy.

To better understand why Gini coefficients can converge to a finite limit even if wealth concentration occurs, consider the following idealized situation:

Example 4.11 Assume $g > 0$ and N numbers $a_1 \leq \cdots \leq a_k < 0$ and $0 \leq a_{k+1} \leq \cdots \leq a_N$ with

$$a = \sum_{i=1}^{N} a_i > 0.$$

For $1 \leq i \leq N$, let the wealth of the ith household group be given by

$$V_t^i = a_i e^{gt}.$$

From this we conclude $\lim_{t\to\infty} V_t^i = -\infty$ for $1 \leq i \leq k$ and $\lim_{t\to\infty} V_t^j = \infty$ for at least one index $k+1 \leq j \leq N$. Thus, wealth concentration occurs in the economy. However,

$$\begin{aligned} G_t &= \frac{N+1}{N} - \frac{2}{N}\frac{1}{V_t}\left(NV_t^1 + (N-1)\,V_t^2 + \cdots + V_t^N\right) \\ &= \frac{N+1}{N} - \frac{2}{N}\frac{1}{a}\left(Na_1 + (N-1)\,a_2 + \cdots + a_N\right), \end{aligned}$$

so G_t is finite and constant over time. With $b_i = a_i/a$ we have $\sum_{i=1}^{N} b_i = 1$, and from Lemma 4.2 or from Corollary 4.6 we obtain

$$G_t = \frac{N+1}{N} - \frac{2}{N}\left(Nb_1 + (N-1)\,b_2 + \cdots + b_N\right) > 0.$$

△

Part II
Discrete-Time Circular Flow Models

Chapter 5
The Discretized Circular Flow Model

The simple differential equations in Sect. 3.3, which describe the evolution of household wealth, result from assuming time-invariant parameters in the circular flow model. If these parameters are allowed to vary over time, the solution methods presented in Sect. 3.3 can no longer be applied.

However, time-varying model parameters are more realistic—especially if household groups are expected to respond to changing circumstances. Suppose a household group is no longer able to finance its planned consumption at a certain point in time because its income is insufficient. In this case, the model should allow for the group to first reduce its consumption to a certain degree before beginning to finance it partially through borrowing.

Another example highlighting the need for time-dependent parameters involves labor income. Suppose the model initially assumes that labor income grows in line with economic growth. It may then turn out that the available aggregate labor income $W = Y - R$ is no longer sufficient at some point to sustain this growth across all household groups. In this case, the model could be adjusted—in line with empirical evidence—to prioritize upper income groups in the distribution of total labor income.

These examples demonstrate that it is both reasonable and necessary for the model to be able to respond to events. Such responses are implemented through the use of rules, introducing nonlinear effects into the modeling. As a result, attempts to find analytical solutions for the dynamic behavior of economic variables become impractical.

To incorporate such rules, the circular flow models are reformulated from continuous to discrete time, enabling a recursive computation of household group development—from one time point to the next. At each time step, specified rules can be used to react to emerging events.

These rules may be designed according to various principles. On the one hand, they can be formulated to reflect observed reality as accurately as possible. On the other hand, one could explore the effects that novel types of rules might have on the

J. Kremer, *Capital Income and Concentration of Wealth*, Springer Studies in Alternative Economics, https://doi.org/10.1007/978-3-658-50715-2_5

development of household groups—potentially leading to deeper insights into the structure and dynamics of economic systems.

5.1 The Recursion Algorithm

In the following, we consider the annual development of an economy for a given period of years. For each year under consideration, the following questions must then be answered using rules to be specified:

- How is the available total labor income distributed among the individual household groups?
- What is the amount of total consumption and how is this amount distributed among the household groups?

These rules cannot be derived from the circular flow diagrams but should be based:

- Either on empirical data
- Or on plausible assumptions

The recursion algorithm can then be implemented using the following flowchart:

- The simulation first requires a set of initial data.
- Then some introductory calculations, the initializations, have to be carried out once.
- Finally, the actual simulation is calculated in a loop for each year from the initial year to the final year under consideration.

Initial Data

The algorithm requires the specification of labor income and consumption expenditure for each household group for the initial year of the simulation as initial data. Furthermore, the wealth of each household group for the year before the starting year must be specified so that the capital income for the starting year can be calculated. We assign the index 0 to the initial year, so that the year before the starting year receives the index -1. Furthermore, the capital return rates and the economic growth rates have to be specified for each year considered in the simulation.

With the symbols used in the book, the required initial data may be described as follows:

$$W_0^i,\ C_0^i,\ V_{-1}^i,\ r_t,\ g_t.$$

Here $i = 1, \ldots, N$, where N is the number of household groups. If T denotes the number of years considered in the simulation, then the annual rate of return on capital r_t has to be specified for $t = 0, \ldots, T$ and the annual economic growth rate g_t for $t = 1, \ldots, T$. Both variables r_t and g_t refer to the period $[t-1, t]$.

Initializations

The following calculations have to be carried out exactly once at the beginning:

- Calculation of the capital income R_0^i that each household group receives in the initial year, i.e.,

$$R_0^i = r_0 V_{-1}^i.$$

  If $V_{-1}^i < 0$ applies to a household group i, then R_0^i denotes the capital costs of this group at the initial point in time.
- Calculation of the amounts S_0^i that each household group can save in the initial year, i.e.,

$$S_0^i = W_0^i + R_0^i - C_0^i.$$

- Calculation of the wealth V_0^i of each household group in the initial year, i.e.,

$$V_0^i = V_{-1}^i + S_0^i.$$

- By summing up all household groups, total capital income R_0, total savings amount S_0, total wealth V_0, total consumption C_0, and total labor income W_0 of the economy can now be calculated.
- Calculation of the gross domestic product Y_0 for the initial year

$$Y_0 = W_0 + R_0.$$

Loop from Year 1 to the Final Year T

- Determination of the new gross domestic product Y_t from the old Y_{t-1} using the growth factor g_t

$$Y_t = (1 + g_t)\, Y_{t-1}. \tag{5.1}$$

- Determination of the capital income R_t^i for each household group

$$R_t^i = r_t V_{t-1}^i. \tag{5.2}$$

The total investment income R_t of the economy can then be calculated by summing the income R_t^i for each household group.

- Calculation of the total income W_t of the economy

$$W_t = Y_t - R_t. \tag{5.3}$$

- Allocation of total labor income W_t to the household groups using the specified rules. We then obtain W_t^i, and thus α_t^i, for each household group i.
- Calculation of total consumption expenditure C_t followed by its allocation to the household groups using the specified rules. We then obtain C_t^i, and thus β_t^i, for each household group i.
- The savings amounts S_t^i can now be calculated for each household group

$$S_t^i = W_t^i + R_t^i - C_t^i. \tag{5.4}$$

By summing up the household groups, we obtain the total savings amount S_t of the economy.

- Finally, the new wealth V_t^i can be calculated using the old V_{t-1}^i and the savings amounts S_t^i

$$V_t^i = V_{t-1}^i + S_t^i. \tag{5.5}$$

By summing up all household groups, we obtain the total wealth V_t of the economy.

- The loop can now be run for the next point in time. The simulation ends when the end point T is reached or if the total labor income of the economy falls to zero.

5.2 Development of Wealth as Dynamical System

This section shows that a discrete-time circular flow economy can be formulated as a linear inhomogeneous dynamical system.

Definition 5.1 (Discrete Dynamical System) Assume that $M \subset \mathbb{R}^n$ is a set and that $f_t : M \to M$ are mappings for $t \in \mathbb{N}$. For some given $x_0 \in M$, the recursion

$$x_t = f_t(x_{t-1}) \quad (t \in \mathbb{N}) \tag{5.6}$$

is called a **discrete dynamical system**. Here, x_0 is the initial value of the recursion, and the series of points $x = (x_0, x_1, \ldots)$ with $x_t = f_t(x_{t-1})$ for $t \in \mathbb{N}$ is the **orbit** of x_0.

Assume that for each t there is a real $n \times n$ matrix $A_t \in \mathbb{R}^{n,n}$ and $b_t \in \mathbb{R}^n$ so that

$$f_t(x_{t-1}) = A_t x_{t-1} + b_t;$$

then the dynamical system is called **linear inhomogeneous**. In the case $b_t = 0$ for all t it is called **linear homogeneous**.

Assume that for each point in time t the available total labor income W_t is distributed to the individual household groups using a rule so that the labor incomes W_t^i of the groups are known for all $i = 1, \ldots, N$. Then the distribution factors α_t^i are given by

$$\alpha_t^i = \frac{W_t^i}{W_t}.$$

Let it be further assumed that for each time t the total consumption C_t and its distribution C_t^i among the household groups have been determined. In this case, the distribution factors β_t^i are given by

$$\beta_t^i = \frac{C_t^i}{C_t}.$$

Then, the following recursive relationship exists for the wealth of the household groups:

Proposition 5.2 *Let V_t^i be the wealth of the ith household group in year t. Then*

$$V_t^i = \sum_{j=1}^{N} a_t^{ij} V_{t-1}^j + b_t^i \tag{5.7}$$

with

$$\begin{aligned} a_t^{ij} &= (1 + r_t)\,\delta^{ij} - r_t \alpha_t^i \\ b_t^i &= \alpha_t^i Y_t - \beta_t^i C_t. \end{aligned} \tag{5.8}$$

Here δ denotes the Kronecker symbol, i.e., $\delta^{ij} = 1$ for $i = j$ and $\delta^{ij} = 0$ for $i \neq j$.

Proof The assertion follows from

$$\begin{aligned} V_t^i &= V_{t-1}^i + S_t^i \\ &= V_{t-1}^i + W_t^i + R_t^i - C_t^i \\ &= V_{t-1}^i + \alpha_t^i\,(Y_t - r_t V_{t-1}) + r_t V_{t-1}^i - C_t^i \\ &= (1 + r_t)\,V_{t-1}^i - r_t \alpha_t^i \sum_{j=1}^{N} V_{t-1}^j + \left(\alpha_t^i Y_t - \beta_t^i C_t\right). \end{aligned}$$

□

The development of the economy's wealth may thus be described as the orbit of a linear inhomogeneous dynamic system with initial value

$$\mathbf{V}_{-1} = \begin{pmatrix} V_{-1}^1 \\ \vdots \\ V_{-1}^N \end{pmatrix}$$

at the initial time $t = -1$. Thus, here the starting time is $t = -1$ and not $t = 0$.

Corollary 5.3 *For a given initial value* $\mathbf{V}_{-1}$,

$$\mathbf{V}_t = \mathbf{A}_t \mathbf{V}_{t-1} + \mathbf{b}_t \tag{5.9}$$

for all $t \in \mathbb{N}_0$. *Here,*

$$\mathbf{V}_t = \begin{pmatrix} V_t^1 \\ \vdots \\ V_t^N \end{pmatrix}, \tag{5.10}$$

and $\mathbf{A}_t \in \mathbb{R}^{N,N}$ *is given by*

$$\begin{aligned} \mathbf{A}_t &= \begin{pmatrix} a_t^{11} & \cdots & a_t^{1N} \\ \vdots & & \vdots \\ a_t^{N1} & \cdots & a_t^{NN} \end{pmatrix} \\ &= \begin{pmatrix} 1 + r_t - r_t\alpha_t^1 & -r_t\alpha_t^1 & \cdots & -r_t\alpha_t^1 \\ -r_t\alpha_t^2 & 1 + r_t - r_t\alpha_t^2 \; -r_t\alpha_t^2 & \cdots & -r_t\alpha_t^2 \\ \vdots & & \ddots & \vdots \\ -r_t\alpha_t^N & \cdots & & -r_t\alpha_t^N \; 1 + r_t - r_t\alpha_t^N \end{pmatrix} \\ &= I + r_t \left(I - \begin{pmatrix} \alpha_t^1 & \cdots & \alpha_t^1 \\ \vdots & & \vdots \\ \alpha_t^N & \cdots & \alpha_t^N \end{pmatrix} \right), \end{aligned} \tag{5.11}$$

where I *denotes the* $N \times N$ *unit matrix. Finally, the vectors* $\mathbf{b}_t$ *are defined by*

$$\mathbf{b}_t = \begin{pmatrix} b_t^1 \\ \vdots \\ b_t^N \end{pmatrix} \tag{5.12}$$

$$= \begin{pmatrix} \alpha_t^1 Y_t - \beta_t^1 C_t \\ \vdots \\ \alpha_t^N Y_t - \beta_t^N C_t \end{pmatrix}.$$

Proof With (5.8), the assertion follows directly from the recursion (5.7). □

Equation (5.9) is the recursion equation for a discrete linear inhomogeneous dynamical system with initial value $\mathbf{V}_{-1}$.

The following chapter shows that, analogous to the continuous-time case, closed-form solutions also exist for the discretized circular flow model with constant parameters. These solutions are similar to those in Chap. 3 and lead to the same conclusions as those that arise for the continuous-time model in Sect. 3.4.

Chapter 6
The Circular Flow Economy with Constant Parameters

If the parameters of the discretized circular flow model are chosen to be constant over time, then closed-form expressions can be derived for the key economic variables.

We assume $S_0 > 0$, $C_0 > 0$, and $V_{-1} > 0$ as well as a constant rate of return on capital $r_t = r > -1$ and a constant economic growth rate $g_t = g > -1$. Thus, for the points in time $t = 0, \ldots, T$, we have

$$Y_t = (1+g)^t \, Y_0 \tag{6.1}$$

and

$$C_t = (1+g)^t \, C_0 \tag{6.2}$$

so that

$$S_t = Y_t - C_t = (1+g)^t \, S_0. \tag{6.3}$$

Furthermore, as before, it is assumed that each household group has a constant share of total labor income W_t and a constant share of total consumption expenditure C_t over time, so that for all $t = 0, \ldots, T$ we may write

$$\begin{aligned} W_t^i &= \alpha^i \, W_t, \\ C_t^i &= \beta^i \, C_t \end{aligned} \tag{6.4}$$

with constant factors $0 \leq \alpha^i \leq 1$, $\sum_{i=1}^N \alpha^i = 1$, and $0 \leq \beta^i \leq 1$, $\sum_{i=1}^N \beta^i = 1$.

J. Kremer, *Capital Income and Concentration of Wealth*, Springer Studies in Alternative Economics, https://doi.org/10.1007/978-3-658-50715-2_6

6.1 Development of the Overall Economy over Time

Proposition 6.1 *In a circular flow economy with constant parameters, total wealth V_t is for $t = -1, \ldots, T$ given by*

$$V_t = \begin{cases} (t+1)\, S_0 + V_{-1} & (g = 0) \\ (1+g)^{t+1} \frac{S_0}{g} + \left(V_{-1} - \frac{S_0}{g}\right) & (g \neq 0)\,. \end{cases} \tag{6.5}$$

Proof The recursive relationship (5.5) leads inductively with (6.3) to

$$V_t = V_{-1} + S_0 \sum_{j=0}^{t} (1+g)^j = \begin{cases} V_{-1} + (t+1)\, S_0 & (g = 0) \\ V_{-1} + \frac{(1+g)^{t+1}-1}{g} S_0 & (g \neq 0)\,, \end{cases}$$

where the summation formula for the geometric sum was used for $g \neq 0$.[1] □

Proposition 6.2 *In a circular flow economy with constant parameters, the aggregated labor income W_t is for $t = 0, \ldots, T$ given by*

$$W_t = \begin{cases} -tr\, S_0 + W_0 & (g = 0) \\ (1+g)^t \left(Y_0 - \frac{r}{g} S_0\right) + \left(\frac{r}{g} S_0 - r\, V_{-1}\right) & (g \neq 0)\,. \end{cases} \tag{6.6}$$

Proof With (5.3), (5.2), and (6.5) we obtain

$$\begin{aligned} W_t &= Y_t - R_t \\ &= Y_t - r\, V_{t-1} \\ &= \begin{cases} Y_0 - tr\, S_0 - r\, V_{-1} & (g = 0) \\ (1+g)^t\, Y_0 - (1+g)^t \frac{r}{g} S_0 + \left(\frac{r}{g} S_0 - r\, V_{-1}\right) & (g \neq 0)\,. \end{cases} \end{aligned}$$

□

[1] We consider the geometric sum $s_n = \sum_{j=0}^{n} q^j$. Then $q s_n - s_n = q^{n+1} - 1$, and for $q \neq 1$ we conclude $s_n = \frac{q^{n+1}-1}{q-1}$. This expression for s_n is called the summation formula for the geometric sum. For $q = 1$, we obviously have $s_n = n + 1$.

6.2 Development of the Household Groups over Time

Now we consider the individual household groups. By definition, we have for $i = 1, \ldots, N$

$$\begin{aligned} S_t^i &= W_t^i + R_t^i - C_t^i \\ &= \alpha^i W_t + r V_{t-1}^i - \beta^i C_t \end{aligned} \tag{6.7}$$

and

$$V_t^i = V_{t-1}^i + S_t^i. \tag{6.8}$$

Proposition 6.3 *For the labor income W_t^i and the consumption expenditure C_t^i of a household group $i = 1, \ldots, N$, we have for $t = 0, \ldots, T$*

$$W_t^i = \begin{cases} -tr\alpha^i S_0 + W_0^i & (g = 0) \\ (1+g)^t \alpha^i \left(Y_0 - \frac{r}{g} S_0 \right) + \alpha^i \left(\frac{r}{g} S_0 - r V_{-1} \right) & (g \neq 0) \end{cases} \tag{6.9}$$

and

$$C_t^i = (1+g)^t C_0^i. \tag{6.10}$$

Proof This follows with Proposition 6.2 and with (6.2) from (6.4). □

Proposition 6.4 *In a circular flow economy with constant parameters, the wealth V_t^i of the ith household group is for $t = -1, \ldots, T$ given by*

$$V_t^i = \begin{cases} (t+1) S_0^i + V_{-1}^i & (r = 0,\ g = 0) \\ \frac{\gamma^{t+1} - 1}{g} S_0^i + V_{-1}^i & (r = 0,\ g \neq 0) \end{cases} \tag{6.11}$$

and

$$V_t^i = \begin{cases} \frac{\rho^{t+1}-1}{r} \left(W_0^i - C_0^i \right) - \left(\sum_{j=0}^{t-1} (t-j) \rho^j \right) \alpha^i r S_0 & (r \neq 0,\ g = 0) \\ \quad + \rho^{t+1} V_{-1}^i & \\ \frac{\rho^{t+1}-1}{r} \left(W_0^i - C_0^i \right) + \left(\sum_{j=0}^{t} (\gamma^{t-j} - 1) \rho^j \right) D^i & (r \neq 0,\ g \neq 0), \\ \quad + \rho^{t+1} V_{-1}^i & \end{cases} \tag{6.12}$$

where

$$\rho = 1 + r, \quad \gamma = 1 + g \tag{6.13}$$

and

$$D^i = \alpha^i \left(1 - \frac{r}{g}\right) S_0 + \left(\alpha^i - \beta^i\right) C_0 \tag{6.14}$$

for all $i = 1, \ldots, N$.

Proof With (6.7) and (6.8) we obtain

$$V_t^i = W_t^i - C_t^i + (1 + r) V_{t-1}^i. \tag{6.15}$$

The resolution of the recursion (6.8) leads to

$$V_t^i = V_{-1}^i + \sum_{j=0}^{t} S_j^i. \tag{6.16}$$

If $r = 0$, we may use Proposition 6.3 with (6.7) to write

$$\begin{aligned} S_t^i &= W_t^i - C_t^i \\ &= (1 + g)^t \left(W_0^i - C_0^i\right) \\ &= (1 + g)^t S_0^i. \end{aligned}$$

Combining this with (6.16) results in the two equations in (6.11).

If $g = 0$, then from (6.15) we conclude with Proposition 6.3

$$V_t^i = \left(W_0^i - C_0^i\right) - t\alpha^i r S_0 + (1 + r) V_{t-1}^i.$$

The resolution of the recursion leads to

$$V_t^i = \left(W_0^i - C_0^i\right) \sum_{j=0}^{t} \rho^j - \alpha^i r S_0 \sum_{j=0}^{t-1} (t - j) \rho^j + \rho^{t+1} V_{-1}^i. \tag{6.17}$$

For $r \neq 0$ this coincides with the first line of (6.12) and specializes to the first line of (6.11) for $r = 0$.

If $g \neq 0$, then from (6.15) and Proposition 6.3 we derive the recursion

$$\begin{aligned} V_t^i &= (1 + g)^t \left(\alpha^i \left(Y_0 - \frac{r}{g} S_0\right) - \beta^i C_0\right) + \alpha^i \left(\frac{r}{g} S_0 - r V_{-1}\right) + (1 + r) V_{t-1}^i \\ &= \gamma^t D^i + A^i + \rho V_{t-1}^i \end{aligned}$$

with

$$D^i = \alpha^i \left(Y_0 - \frac{r}{g} S_0\right) - \beta^i C_0$$
$$= \alpha^i \left(1 - \frac{r}{g}\right) S_0 + \left(\alpha^i - \beta^i\right) C_0,$$

where $Y_0 = C_0 + S_0$ has been used, and with

$$A^i = \alpha^i \left(\frac{r}{g} S_0 - r V_{-1}\right).$$

Inductively we find

$$V_t^i = A^i \sum_{j=0}^{t} \rho^j + D^i \sum_{j=0}^{t} \gamma^{t-j} \rho^j + \rho^{t+1} V_{-1}^i \tag{6.18}$$
$$= \frac{\rho^{t+1} - 1}{r} \left(W_0^i - C_0^i\right) + D^i \sum_{j=0}^{t} \left(\gamma^{t-j} - 1\right) \rho^j + \rho^{t+1} V_{-1}^i$$

where we used

$$A^i + D^i = \alpha^i \left(\frac{r}{g} S_0 - r V_{-1}\right) + \alpha^i \left(1 - \frac{r}{g}\right) S_0 + \left(\alpha^i - \beta^i\right) C_0 \tag{6.19}$$
$$= \alpha^i (S_0 - R_0) + \left(\alpha^i - \beta^i\right) C_0$$
$$= \alpha^i (W_0 - C_0) + \left(\alpha^i - \beta^i\right) C_0$$
$$= \alpha^i W_0 - \beta^i C_0$$
$$= W_0^i - C_0^i$$

for the second line of (6.18). □

The two equations in (6.11) are analogous to the equations in (6.5) for the overall economy: If, in the case $r = 0$, the initial income W_0^i exceeds the initial consumption C_0^i of the ith household group, i.e., $S_0^i > 0$, then debt never occurs, and the group participates in economic growth at all times.

Lemma 6.5 *For $r \neq 0$ and $g \neq r$, we have*

$$\sum_{j=0}^{t} \left(\gamma^{t-j} - 1\right) \rho^j = \frac{1}{r} \frac{r\gamma^{t+1} - g\rho^{t+1}}{g - r} + \frac{1}{r}. \tag{6.20}$$

Proof With (6.13) we compute

$$\begin{aligned}\sum_{j=0}^{t}\left(\gamma^{t-j}-1\right)\rho^{j}&=\gamma^{t}\sum_{j=0}^{t}\left(\frac{\rho}{\gamma}\right)^{j}-\sum_{j=0}^{t}\rho^{j}\\&=\gamma^{t}\frac{\left(\frac{\rho}{\gamma}\right)^{t+1}-1}{\frac{\rho}{\gamma}-1}-\frac{\rho^{t+1}-1}{\rho-1}\\&=\frac{\rho^{t+1}-\gamma^{t+1}}{\rho-\gamma}-\frac{\rho^{t+1}-1}{\rho-1}\\&=\frac{r\rho^{t+1}-r\gamma^{t+1}-(r-g)\,\rho^{t+1}}{r\,(r-g)}+\frac{1}{r},\end{aligned}$$

as was to be shown. □

Lemma 6.6 *For $r \neq 0$, we have*

$$\sum_{j=0}^{t} j\rho^{j}=\frac{t\rho^{t+2}-(t+1)\,\rho^{t+1}+\rho}{(\rho-1)^{2}}. \tag{6.21}$$

Proof For $t = 0$, the left side of (6.21) coincides with the right side, both being equal to zero. Assume that (6.21) has already been proved for some t. Then

$$\begin{aligned}\sum_{j=0}^{t+1} j\rho^{j}&=\sum_{j=0}^{t} j\rho^{j}+(t+1)\,\rho^{t+1}\\&=\frac{t\rho^{t+2}-(t+1)\,\rho^{t+1}+\rho}{(\rho-1)^{2}}+\frac{(\rho-1)^{2}\,(t+1)\,\rho^{t+1}}{(\rho-1)^{2}}\\&=\frac{(t+1)\,\rho^{t+3}+(t-2t-2)\,\rho^{t+2}+\rho}{(\rho-1)^{2}},\end{aligned}$$

what was to be shown. □

Alternatively, (6.21) can also be proved with $s_n = \sum_{j=0}^{n} q^j = \frac{q^{n+1}-1}{q-1}$ by calculating $q\frac{\mathrm{d}}{\mathrm{d}q}s_n$. Using the previous two lemmas, the formulas in (6.12) for $r \neq 0$ are now simplified.

Proposition 6.7 *In a circular flow economy with constant parameters, we have for $i = 1, \ldots, N$ and for $t = -1, \ldots, T$*

$$V_t^i = \begin{cases} \rho^{t+1}\frac{1}{r}\Delta^i + (t+1)\,\alpha^i S_0 - \frac{1}{r}\Delta^i + V_{-1}^i & (r \neq 0,\ g = 0) \\ \rho^{t+1}\left(\frac{1}{r-g}\Delta_1^i + \Delta_2^i\right) + \gamma^{t+1}\left(\frac{1}{g-r}\Delta_1^i + \frac{\alpha^i}{g}S_0\right) + \xi^i & (r \neq 0,\ g \neq 0, \\ & \ r \neq g) \\ \rho^{t+1}\left(\left((t+1)\frac{1}{\rho} - \frac{1}{r}\right)\Delta_1^i + \frac{1}{r}S_0^i\right) + \xi^i & (r \neq 0,\ g \neq 0, \\ & \ r = g) \end{cases} \tag{6.22}$$

with

$$\xi^i = \alpha^i \left(V_{-1} - \frac{S_0}{g}\right) \tag{6.23}$$

and

$$\begin{aligned} \Delta_1^i &= \left(\alpha^i - \beta^i\right) C_0 \\ \Delta_2^i &= V_{-1}^i - \alpha^i V_{-1} \\ \Delta^i &= S_0^i - \alpha^i S_0 \\ &= \left(\alpha^i - \beta^i\right) C_0 + r\left(V_{-1}^i - \alpha^i V_{-1}\right) \\ &= \Delta_1^i + r\Delta_2^i. \end{aligned} \tag{6.24}$$

Proof First, $g = 0$ is assumed. With Lemma 6.6 we write

$$\begin{aligned} \sum_{j=0}^{t-1} (t-j)\,\rho^j &= t\frac{\rho^t - 1}{r} - \frac{(t-1)\,\rho^{t+1} - t\rho^t + \rho}{r^2} \\ &= \frac{\rho^{t+1} - \rho - tr}{r^2}. \end{aligned}$$

If this is inserted into the first line of (6.12), we obtain

$$\begin{aligned} V_t^i &= \frac{\rho^{t+1} - 1}{r}\left(W_0^i - C_0^i\right) - \frac{\rho^{t+1} - \rho - tr}{r}\alpha^i S_0 + \rho^{t+1} V_{-1}^i \\ &= \frac{\rho^{t+1}}{r}\left(W_0^i - C_0^i - \alpha^i S_0 + rV_{-1}^i\right) + t\alpha^i S_0 - \frac{1}{r}\left(W_0^i - C_0^i - \rho\alpha^i S_0\right). \end{aligned} \tag{6.25}$$

With $S_t^i = W_t^i + R_t^i - C_t^i$ and $R_0^i = rV_{-1}^i$, we write

$$W_0^i - C_0^i - \alpha^i S_0 + rV_{-1}^i = S_0^i - \alpha^i S_0 = \Delta^i.$$

Further, we have

$$\begin{aligned} W_0^i - C_0^i - \rho\alpha^i S_0 &= S_0^i - R_0^i - (1+r)\alpha^i S_0 \\ &= \left(S_0^i - \alpha^i S_0\right) - rV_{-1}^i + r\alpha^i S_0, \end{aligned}$$

leading to the first line of (6.22).

For $r \neq 0$, $g \neq 0$, and $r \neq g$ using Lemma 6.5 and (6.14), the second line of (6.12) may be written as

$$\begin{aligned} V_t^i &= \frac{\rho^{t+1} - 1}{r}\left(W_0^i - C_0^i\right) + \left(\frac{1}{r}\frac{r\gamma^{t+1} - g\rho^{t+1}}{g-r} + \frac{1}{r}\right) D^i + \rho^{t+1} V_{-1}^i \\ &= \frac{\rho^{t+1}}{r}\left(W_0^i - C_0^i - \frac{gD^i}{g-r} + rV_{-1}^i\right) + \gamma^{t+1}\frac{D^i}{g-r} + \frac{D^i - W_0^i + C_0^i}{r}. \end{aligned}$$

With $D^i = \alpha^i \frac{g-r}{g} S_0 + \Delta_1^i$ we have

$$\begin{aligned} W_0^i - C_0^i - \frac{g}{g-r}D^i + rV_{-1}^i &= S_0^i - \alpha^i S_0 - \frac{g}{g-r}\Delta_1^i \\ &= \Delta_1^i + r\Delta_2^i - \frac{g}{g-r}\Delta_1^i \\ &= -\frac{r}{g-r}\Delta_1^i + r\Delta_2^i \end{aligned}$$

and

$$\frac{D^i}{g-r} = \frac{1}{g-r}\Delta_1^i + \frac{\alpha^i}{g}S_0.$$

With (6.19) we obtain

$$D^i - W_0^i + C_0^i = -A^i = \alpha^i r\left(V_{-1} - \frac{S_0}{g}\right),$$

and thus the second line of (6.22).

If $r \neq 0$, $g \neq 0$, and $r = g$, (6.14) reduces to

$$D^i = \left(\alpha^i - \beta^i\right) C_0 = \Delta_1^i.$$

Because $\rho = \gamma$ we have

$$\sum_{j=0}^{t}\left(\gamma^{t-j}-1\right)\rho^{j}=(t+1)\,\rho^{t}-\frac{\rho^{t+1}-1}{r},$$

and the second line of (6.12) may therefore be rewritten as

$$\begin{aligned}
V_t^i &= \frac{\rho^{t+1}-1}{r}\left(W_0^i-C_0^i\right)+\left(\sum_{j=0}^{t}\left(\gamma^{t-j}-1\right)\rho^{j}\right)D^i+\rho^{t+1}V_{-1}^i \qquad (6.26)\\
&= \frac{\rho^{t+1}-1}{r}\left(W_0^i-C_0^i\right)+\left((t+1)\,\rho^{t}-\frac{\rho^{t+1}-1}{r}\right)\Delta_1^i+\rho^{t+1}V_{-1}^i\\
&= \frac{\rho^{t+1}-1}{r}\left(W_0^i-C_0^i-\Delta_1^i\right)+(t+1)\,\rho^{t}\Delta_1^i+\rho^{t+1}V_{-1}^i\\
&= \rho^{t+1}\frac{1}{r}\left(W_0^i-C_0^i+rV_{-1}^i-\Delta_1^i\right)+(t+1)\,\rho^{t}\Delta_1^i-\frac{1}{r}\left(W_0^i-C_0^i-\Delta_1^i\right).
\end{aligned}$$

With $R_0^i = rV_{-1}^i$ and $S+C=W+R$, we conclude

$$\begin{aligned}
W_0^i-C_0^i-\Delta_1^i &= \alpha^i W_0 - C_0^i + C_0^i - \alpha^i C_0 \qquad (6.27)\\
&= \alpha^i\,(W_0-C_0)\\
&= \alpha^i\,(S_0-R_0)\\
&= \alpha^i\,(S_0-rV_{-1})
\end{aligned}$$

and

$$W_0^i-C_0^i+rV_{-1}^i=S_0^i. \qquad (6.28)$$

If (6.27) and (6.28) are inserted into (6.26), we get with (6.23)

$$V_t^i=\rho^{t+1}\frac{1}{r}\left(S_0^i-\Delta_1^i\right)+(t+1)\,\rho^{t}\Delta_1^i+\xi^i,$$

as was to be shown. □

Analogous to (3.22) we define

$$\upsilon^i=\frac{V_{-1}^i}{V_{-1}}, \qquad (6.29)$$

this being well defined because we have $V_{-1}>0$ by assumption. The results of this section are now summarized as follows:

Proposition 6.8 *In a circular flow economy with constant parameters, the wealth V_t^i of the ith household group is for $t = -1, \ldots, T$ given by*

$$V_t^i = \begin{cases} tS_0^i + V_{-1}^i + S_0^i & (r=0,\ g=0) \\ \gamma^{t+1}\frac{1}{g}S_0^i + \left(V_{-1}^i - \frac{1}{g}S_0^i\right) & (r=0,\ g\neq 0) \\ \rho^{t+1}\left(\frac{1}{r}\Delta_1^i + \Delta_2^i\right) + (t+1)\,\alpha^i S_0 & (r\neq 0,\ g=0) \\ \quad +\left(V_{-1}^i - \left(\frac{1}{r}\Delta_1^i + \Delta_2^i\right)\right) & \\ \rho^{t+1}\left(\left((t+1)\frac{1}{\rho} - \frac{1}{r}\right)\Delta_1^i + \frac{1}{r}S_0^i\right) + \xi^i & (r\neq 0,\ g\neq 0,\ r=g) \\ \rho^{t+1}\left(\frac{1}{r-g}\Delta_1^i + \Delta_2^i\right) & (r\neq 0,\ g\neq 0,\ r\neq g) \\ \quad +\gamma^{t+1}\left(\frac{1}{g-r}\Delta_1^i + \frac{\alpha^i}{g}S_0\right) + \xi^i & \end{cases} \tag{6.30}$$

with

$$\xi^i = \alpha^i\left(V_{-1} - \frac{S_0}{g}\right) \tag{6.31}$$

and

$$\Delta_1^i = \left(\alpha^i - \beta^i\right)C_0 \tag{6.32}$$
$$\Delta_2^i = \left(\upsilon^i - \alpha^i\right)V_{-1}.$$

□

6.3 Concentration of Wealth

The circular flow model of this chapter is a discrete-time version of the continuous-time model from Chap. 3, and the formulas in Proposition 6.8 are similar to those in Proposition 3.9. Consider, for example, the case $r \neq 0,\ g \neq 0,\ r \neq g$, for which we have, according to Proposition 3.9, in the continuous-time framework

$$V_t^i = e^{rt}\left(\frac{1}{r-g}\Delta_1^i + \Delta_2^i\right) + e^{gt}\left(\frac{1}{g-r}\Delta_1^i + \frac{\alpha^i}{g}S_0\right) + \xi^i.$$

The corresponding relationship in Proposition 6.8 reads with $\rho = 1 + r$ and with $\gamma = 1 + g$

$$V_t^i = \rho^{t+1}\left(\frac{1}{r-g}\Delta_1^i + \Delta_2^i\right) + \gamma^{t+1}\left(\frac{1}{g-r}\Delta_1^i + \frac{\alpha^i}{g}S_0\right) + \xi^i.$$

The constants ξ^i and Δ_2^i are formed in (3.31) and (3.32) using V_0^i, whereas in (6.31) and (6.32) they are formed using V_{-1}^i, but the conclusions from Chap. 3 are also valid in the discretized circular flow model. In particular, if Definition 3.14 of wealth concentration is adopted in the discrete-time context, the statement of Proposition 3.14 also applies to the discretized version.

6.4 Implementation of Wealth Development

Assume that the rates of return on capital and the economic growth rates in Sect. 5.2 are chosen to be constant over time

$$r_t = r, \quad g_t = g.$$

Furthermore, time-constant allocation factors for the distribution of aggregate labor income and total consumption among the household groups are presupposed, i.e.,

$$\alpha_t^i = \alpha^i, \quad \beta_t^i = \beta^i.$$

In this case, the coefficients a_t^{ij} of the recurrence matrices $\mathbf{A}_t$ in (5.8) are also constant in time and read

$$a_t^{ij} = a^{ij} = \delta_{ij} + r\left(\delta_{ij} - \alpha^i\right),$$

so that

$$\mathbf{A}_t = \mathbf{A} = \begin{pmatrix} 1 + r\left(1 - \alpha^1\right) & -r\alpha^1 & \cdots & -r\alpha^1 \\ -r\alpha^2 & 1 + r\left(1 - \alpha^2\right) & \cdots & -r\alpha^2 \\ \vdots & & \ddots & \vdots \\ -r\alpha^N & & \cdots & 1 + r\left(1 - \alpha^N\right) \end{pmatrix}$$

$$= I + r\left(I - \begin{pmatrix} \alpha^1 & \cdots & \alpha^1 \\ \vdots & & \vdots \\ \alpha^N & \cdots & \alpha^N \end{pmatrix}\right).$$

The b_t^i in (5.8) simplify to

$$b_t^i = \left(\alpha^i Y_0 - \beta^i C_0\right)(1 + g)^t,$$

so that for the development of wealth in an economy with constant parameters we get the dynamical system

$$\mathbf{V}_t = \mathbf{A}\mathbf{V}_{t-1} + \mathbf{b}_t \tag{6.33}$$

with

$$\mathbf{b}_t = \mathbf{b}_0 (1+g)^t , \quad \mathbf{b}_0 = \begin{pmatrix} \alpha^1 Y_0 - \beta^1 C_0 \\ \vdots \\ \alpha^N Y_0 - \beta^N C_0 \end{pmatrix}$$

and with initial value

$$\mathbf{V}_{-1} = \begin{pmatrix} V_{-1}^1 \\ \vdots \\ V_{-1}^N \end{pmatrix} .$$

Remark The recursion (6.33) of the dynamical system can be resolved because

$$\begin{aligned}
\mathbf{V}_0 &= \mathbf{A}\mathbf{V}_{-1} + \mathbf{b}_0 \\
\mathbf{V}_1 &= \mathbf{A}\mathbf{V}_0 + \mathbf{b}_1 \\
&= \mathbf{A}^2\mathbf{V}_{-1} + \mathbf{A}\mathbf{b}_0 + \mathbf{b}_1 \\
\mathbf{V}_2 &= \mathbf{A}\mathbf{V}_1 + \mathbf{b}_2 \\
&= \mathbf{A}^3\mathbf{V}_{-1} + \mathbf{A}^2\mathbf{b}_0 + \mathbf{A}\mathbf{b}_1 + \mathbf{b}_2 \\
&\vdots
\end{aligned}$$

so that

$$\begin{aligned}
\mathbf{V}_t &= \mathbf{A}^{t+1}\mathbf{V}_{-1} + \sum_{j=0}^{t} \mathbf{A}^j \mathbf{b}_{t-j} && (6.34) \\
&= \mathbf{A}^{t+1}\mathbf{V}_{-1} + \sum_{j=0}^{t} \left(\mathbf{A}^j \mathbf{b}_0\right)(1+g)^{t-j} .
\end{aligned}$$

An alternative to using (6.33) or (6.34) to calculate the evolution of wealth is to formulate the equations for V_t^i in Proposition 6.8 in vector form. We use $\mathbb{R}^{m,n}$ to denote the set of real-valued $m \times n$ matrices and define the following row vectors from $\mathbb{R}^{1,N}$:

$$\boldsymbol{V_{-1}} = \left(V_{-1}^1, \ldots, V_{-1}^N\right)$$

$$\boldsymbol{S_0} = \left(S_0^1, \ldots, S_0^N\right)$$

$$\boldsymbol{\alpha} = \left(\alpha^1, \ldots, \alpha^N\right)$$

$$\boldsymbol{\beta} = \left(\beta^1, \ldots, \beta^N\right)$$

$$\boldsymbol{v} = \boldsymbol{V_{-1}}/V_{-1}$$

$$\boldsymbol{\Delta_1} = \left(\Delta_1^1, \ldots, \Delta_1^N\right)$$

$$\boldsymbol{\Delta_2} = \left(\Delta_2^1, \ldots, \Delta_2^N\right).$$

Further, we define the column vector

$$\boldsymbol{e} = \begin{pmatrix} 1 \\ \vdots \\ 1 \end{pmatrix} \in \mathbb{R}^{T+2,1}$$

whose entries consist only of ones. If $\boldsymbol{a} = \left(a^1, \ldots, a^N\right)$ is a row vector, then

$$\boldsymbol{ea} = \begin{pmatrix} 1 \\ \vdots \\ 1 \end{pmatrix} \left(a^1, \ldots, a^N\right) = \begin{pmatrix} a^1 & \cdots & a^N \\ \vdots & & \vdots \\ a^1 & \cdots & a^N \end{pmatrix} \in \mathbb{R}^{T+2,N}.$$

Finally, we define the column vectors

$$\boldsymbol{t} = \begin{pmatrix} -1 \\ 0 \\ \vdots \\ T \end{pmatrix} \in \mathbb{R}^{T+2,1}, \quad \boldsymbol{\rho} = \begin{pmatrix} 1 \\ \rho \\ \rho^2 \\ \vdots \\ \rho^{T+1} \end{pmatrix} \in \mathbb{R}^{T+2,1}, \quad \boldsymbol{\gamma} = \begin{pmatrix} 1 \\ \gamma \\ \gamma^2 \\ \vdots \\ \gamma^{T+1} \end{pmatrix} \in \mathbb{R}^{T+2,1}.$$

Proposition 6.8 can now be formulated as follows:

Proposition 6.9 *In a circular flow economy with constant parameters, the wealth*

$$\boldsymbol{V} = \begin{pmatrix} V_{-1}^1 & \cdots & V_{-1}^N \\ \vdots & & \vdots \\ V_T^1 & \cdots & V_T^N \end{pmatrix} \in \mathbb{R}^{T+2,N}$$

of the household groups $i = 1, \ldots, N$ for the points in time $t = -1, \ldots, T$ is given by

$$V = \begin{cases} t\boldsymbol{S}_0 + \boldsymbol{e}\,(\boldsymbol{V}_{-1} + \boldsymbol{S}_0) & (r = 0,\ g = 0) \\ \frac{1}{g}\boldsymbol{\gamma}\,\boldsymbol{S}_0 + \boldsymbol{e}\left(\boldsymbol{V}_{-1} - \frac{1}{g}\boldsymbol{S}_0\right) & (r = 0,\ g \neq 0) \\ \boldsymbol{\rho\Delta} + S_0\,(\boldsymbol{t} + \boldsymbol{e})\,\boldsymbol{\alpha} + \boldsymbol{e}\,(\boldsymbol{V}_{-1} - \boldsymbol{\Delta}) & (r \neq 0,\ g = 0) \\ \operatorname{diag}(\boldsymbol{\rho})\left(\frac{1}{\rho}\boldsymbol{t}\boldsymbol{\Delta}_1 + \left(\frac{1}{\rho} - \frac{1}{r}\right)\boldsymbol{e}\boldsymbol{\Delta}_1 + \frac{1}{r}\boldsymbol{e}\boldsymbol{S}_0\right) + \boldsymbol{e\xi} & (r \neq 0,\ g \neq 0,\ r = g) \\ \boldsymbol{\rho}\left(\frac{1}{r-g}\boldsymbol{\Delta}_1 + \boldsymbol{\Delta}_2\right) + \boldsymbol{\gamma}\left(\frac{1}{g-r}\boldsymbol{\Delta}_1 + \frac{S_0}{g}\boldsymbol{\alpha}\right) + \boldsymbol{e\xi} & (r \neq 0,\ g \neq 0,\ r \neq g) \end{cases} \tag{6.35}$$

with

$$\boldsymbol{\xi} = \left(\boldsymbol{V}_{-1} - \frac{S_0}{g}\right)\boldsymbol{\alpha}$$

and

$$\boldsymbol{\Delta} = \frac{1}{r}\boldsymbol{\Delta}_1 + \boldsymbol{\Delta}_2,$$

as well as

$$\operatorname{diag}(\boldsymbol{\rho}) = \begin{pmatrix} \rho^0 & \cdots & 0 \\ \vdots & \ddots & \vdots \\ 0 & \cdots & \rho^{T+1} \end{pmatrix} \in \mathbb{R}^{T+2,T+2}.$$

□

In a matrix-oriented programming language, such as MATLAB, the formulas (6.33), (6.34), and also (6.35) can be programmed efficiently.

Chapter 7
Rules for the Discretized Circular Flow Model

Until now, time-constant parameters have been assumed. This guarantees the existence of analytical expressions for the temporal development of important economic variables, such as labor income, wealth or consumer spending, but are not found in real economies.

The purpose of the transition from continuous-time to discrete-time models is to be able to react to events that occur over time with the help of rules. This leads to nonlinear effects in the simulations, and we no longer intend to derive formulas, e.g., for the wealth of a household group at a specific point in time. The aim is rather to study the influence of different rules on the development of economies, in particular on wealth concentration, using simulations.

7.1 Rules for the Allocation of Labor Income to the Household Groups

To allocate the aggregate labor income W_t of the economy to the household groups, $W_t^i = \alpha_t^i W_t$, we present two modeling approaches. Since an initial income $W_0^i > 0$ is given for each household group, we have

$$\alpha_0^i = \frac{W_0^i}{W_0} > 0$$

for all $i = 1, \ldots, N$.

Rule: The share of labor income of each household group in total labor income is constant over time

Here we define

J. Kremer, *Capital Income and Concentration of Wealth*, Springer Studies in Alternative Economics, https://doi.org/10.1007/978-3-658-50715-2_7

$$\frac{W_t^i}{W_t} = \alpha_t^i = \alpha_0^i = \frac{W_0^i}{W_0} \tag{7.1}$$

for each group $i = 1, \ldots, N$ and for all $t = 0, \ldots, T$. This is the rule that applies in a circular flow economy with constant parameters. It is simple but unrealistic, because if the available total labor income decreases over time, then in the case of constant parameters the labor income of all household groups would decrease over time, which is not observed in reality.

On the other hand, this rule limits the spread of labor income because we obtain from (7.1) for $1 \leq i < j \leq N$

$$\frac{W_t^j}{W_t^i} = \frac{W_t^j}{W_t}\frac{W_t}{W_t^i} = \frac{\alpha_0^j}{\alpha_0^i}. \tag{7.2}$$

Rule: The growth in labor income of the household groups is based on the growth in gross domestic product, but the earned income of household groups $1, \ldots, N-1$ is reduced if this cannot be financed

Figure 7.1 shows that the disposable personal income of the top tenth decile of the German economy has risen comparatively sharply in the past. In contrast, the disposable personal income of the lower first decile has fallen slightly over time.

The rule given below can result in different time dependent income growth rates for the different household groups.

According to (2.8), the aggregated labor income of the economy is given by

$$W_t = Y_t - R_t.$$

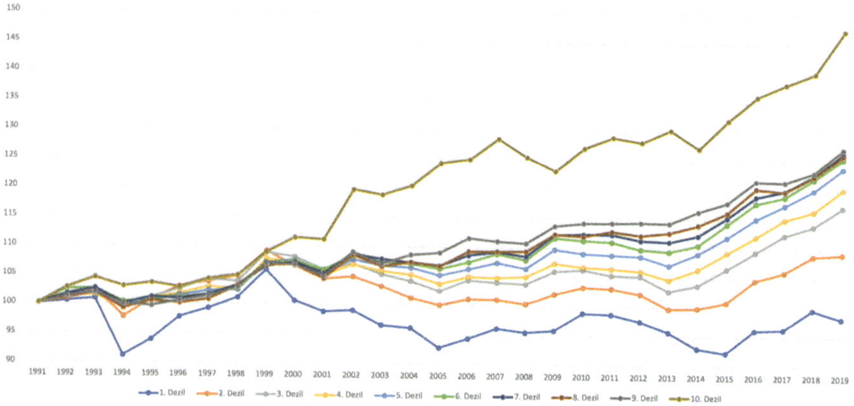

Fig. 7.1 Development of disposable personal income in the German economy by decile, with 1991 identified as 100%. Reprinted with the kind permission of Dr. Markus Grabka, DIW Berlin

If there is zero capital income, $R_t = 0$, then $Y_t = W_t$, and the increases in labor income correspond exactly to economic growth. However, if R_t grows faster than the gross domestic product, then labor income cannot increase with gross domestic product for all household groups. But the rule given below aims to at least allow the labor income of the highest income group N to increase with economic growth. The rule further specifies how the labor income of the lower groups $1, \ldots, N-1$ is to be reduced.

We fix a point in time $t > 0$ and assume $W_{t-1}^i \geq 0$ for all $i = 1, \ldots, N$ and $W_{t-1} > 0$. First, using the economic growth rate, we define a preliminary labor income $\tilde{W}_t^i$ for each household group by

$$\tilde{W}_t^i = (1 + g_t)\, W_{t-1}^i$$

for $i = 1, \ldots, N$. We denote $\tilde{W}_t^i$ the **projected labor income** of household group i at time t. With

$$\tilde{W}_t = \sum_{i=1}^{N} \tilde{W}_t^i$$

we denote the **projected total labor income** of the economy at time t. We assume that the household groups are numbered in ascending order of income, so that the Nth group is the one with the highest income.

We distinguish between the following cases:

First Case

$$W_t = Y_t - R_t \leq 0.$$

In this case, there is no labor income left in the economy that can be distributed, and the simulation is aborted.

Second Case

$$0 < \tilde{W}_t \leq W_t.$$

In this case, the available aggregate labor income W_t exceeds the projected aggregate labor income $\tilde{W}_t$ at time t, and W_t is distributed proportionally to the projected amounts $\tilde{W}_t^i$ to each household group. We therefore define

$$W_t^i = \frac{\tilde{W}_t^i}{\tilde{W}_t} W_t,$$

so that

$$\alpha_t^i = \frac{W_t^i}{W_t} = \frac{\tilde{W}_t^i}{\tilde{W}_t} = \frac{W_{t-1}^i}{W_{t-1}}.$$

In the present case we have in particular $0 < \tilde{W}_t^N \leq W_t$.

Third Case

$$0 < W_t < \tilde{W}_t.$$

In this case, the projected total labor income $\tilde{W}_t$ exceeds the available amount W_t. There are two possibilities:

(a)

$$0 < W_t \leq \tilde{W}_t^N.$$

In this case, the aggregate labor income W_t to be distributed is less than the amount $\tilde{W}_t^N$ that the group with the highest income should receive due to economic growth. We then define

$$W_t^N = W_t \quad \text{and} \quad W_t^1 = \cdots = W_t^{N-1} = 0,$$

so

$$\alpha_t^N = 1 \quad \text{and} \quad \alpha_t^1 = \cdots = \alpha_t^{N-1} = 0.$$

We therefore allocate the entire available labor income W_t to group N with the highest income, while all other groups are left empty-handed.

(b)

$$0 < \tilde{W}_t^N < W_t.$$

Because of $W_t < \tilde{W}_t$, there is at least one other group besides the Nth whose projected labor income is positive. The idea in this case is to set $W_t^N = \tilde{W}_t^N$ and to lower the projected labor income of the remaining groups appropriately so that the available labor income W_t is exhausted exactly.

In order to reduce the labor income of all smaller groups $i < N$ to an amount below their projected value $\tilde{W}_t^i$, we look for a parameter $\lambda > 0$, so that

$$\begin{aligned} W_t &= e^{-\lambda(N-1)}\tilde{W}_t^1 + \cdots + e^{-\lambda}\tilde{W}_t^{N-1} + \tilde{W}_t^N \\ &= e^{-\lambda N}\sum_{i=1}^{N} e^{\lambda i}\tilde{W}_t^i. \end{aligned}$$

The larger the factor λ, the more the labor income for the household groups $i < N$ will be reduced. With the following proposition, we prove both the existence and the uniqueness of this parameter λ.

Proposition 7.1 *Let values* $0 \leq \tilde{W}_t^1 \leq \cdots \leq \tilde{W}_t^N$ *and* W_t *be given with the properties*

$$0 < \tilde{W}_t^N < W_t \tag{7.3}$$

and

$$\sum_{i=1}^{N} \tilde{W}_t^i > W_t. \tag{7.4}$$

Then there is a uniquely defined parameter $\lambda > 0$, *so that*

$$W_t = e^{-\lambda N} \sum_{i=1}^{N} e^{\lambda i} \tilde{W}_t^i.$$

Proof We define a polynomial P by

$$P(x) = x^{N-1} \tilde{W}_t^1 + \cdots + x \tilde{W}_t^{N-1} + \left(\tilde{W}_t^N - W_t \right).$$

Then $P\left(e^{-\lambda}\right) = e^{-\lambda N} \sum_{i=1}^{N} e^{\lambda i} \tilde{W}_t^i - W_t$ and $P(0) = \tilde{W}_t^N - W_t < 0$. Further, (7.3) and (7.4) imply $\tilde{W}_t^{N-1} > 0$, and for $y = \left(W_t - \tilde{W}_t^N \right) / \tilde{W}_t^{N-1} > 0$ we obtain

$$P(y) = y^{N-1} \tilde{W}_t^1 + \cdots + y^2 \tilde{W}_t^{N-2} \geq 0.$$

As a continuous function, P therefore has a root $x_0 \in (0, y]$ according to the intermediate value theorem. Furthermore,

$$P'(x) = (N-1) x^{N-2} \tilde{W}_t^1 + \cdots + 2x \tilde{W}_t^{N-2} + \tilde{W}_t^{N-1} > 0$$

for all $x \geq 0$. So P is strictly monotonically increasing on $(0, \infty)$, and the root x_0 is therefore uniquely determined. Finally, we obtain λ_0 by the condition $x_0 = e^{-\lambda_0}$, which concludes the proof. □

The root $x_0 > 0$ of P can be calculated using a standard method, such as Newton's algorithm. Then, with $x_0 = e^{-\lambda_0}$, we have

$$W_t = \sum_{i=1}^{N} e^{-\lambda_0 (N-i)} \tilde{W}_t^i$$

$$= \sum_{i=1}^{N} \left(e^{-\lambda_0 (N-i)} \frac{\tilde{W}_t^i}{W_t} \right) W_t .$$

Thus, if we define

$$\alpha_t^i = e^{-\lambda_0 (N-i)} \frac{\tilde{W}_t^i}{W_t}$$

and

$$W_t^i = \alpha_t^i W_t ,$$

then we obtain the relationship

$$W_t = \sum_{i=1}^{N} \alpha_t^i W_t = \sum_{i=1}^{N} W_t^i .$$

The α_t^i define the desired distribution of total labor income W_t among the individual household groups.

7.2 Rules on the Growth of Consumer Spending and Its Distribution Among Household Groups

Two rules are given for growth and for the allocation of aggregate consumer spending to the household groups.

Rule: The share of consumption of each household group in total consumption is constant over time

With the reasoning preceding (3.2), it is assumed that the total consumption has the same growth rate as the gross domestic product. Thus,

$$C_t = (1 + g_t)\, C_{t-1} \tag{7.5}$$

for $t = 1, \ldots, T$.

Within the framework of this rule, we further assume that the distribution factors β_t^i are fixed for each group i and for each time t by the initial distribution, so that

$$\beta_t^i = \beta_0^i = \frac{C_0^i}{C_0} \tag{7.6}$$

for all $t = 0, \ldots, T$ and for all $i = 1, \ldots, N$. This is the rule that applies in a circular flow economy with constant parameters. It is simple but problematic if the labor income of a household group falls to such an extent that it can no longer finance its consumption expenditure.

Rule: The growth in consumer spending of a household group is reduced if income is insufficient to finance it

Initially, the consumption expenditures of the individual household groups increase in line with economic growth according to (7.5). However, as part of this rule, it is modeled that the household groups restrict their consumption to a certain extent if their income is no longer sufficient to finance it. Only if income continues to decline will consumer spending have to be financed in part by debt.

For $t > 0$ we define preliminary consumption expenditures

$$\tilde{C}_t^i = (1 + g_t)\, C_{t-1}^i \tag{7.7}$$

of the household groups, which are referred to as **projected consumption expenditures**. C_t^i, on the other hand, denotes as usual the **realized consumption expenditure** at time t, which is defined as follows: First, it is checked for each household group i whether its income $Y_t^i = W_t^i + R_t^i$ is sufficient to finance the projected consumption $\tilde{C}_t^i$. If it suffices, i.e., if $\tilde{C}_t^i \leq Y_t^i$, then the projected consumption expenditure becomes the realized consumption expenditure C_t^i, and thus

$$C_t^i = \tilde{C}_t^i$$

is defined.

However, if the projected consumption $\tilde{C}_t^i$ cannot be fully financed, it is assumed that households restrict their consumption up to the level of so-called **critical consumption expenditure** $\hat{C}_t^i \leq \tilde{C}_t^i$. If Y_t^i is between $\hat{C}_t^i$ and $\tilde{C}_t^i$, then we define

$$C_t^i = Y_t^i.$$

Consumption is therefore initially reduced to the amount that is financeable. Only if Y_t^i falls below $\hat{C}_t^i$, the realized consumption is set to $C_t^i = \hat{C}_t^i$, and the corresponding household group must take on debt to finance this consumption.

The critical consumption expenditure may be modeled with the help of **critical growth factors** $\hat{g}_t \leq g_t$ so that

$$\hat{C}_t^i = \left(1 + \hat{g}_t\right) C_{t-1}^i.$$

To summarize, we therefore define

$$C_t^i = \begin{cases} \tilde{C}_t^i & \text{if } \tilde{C}_t^i \leq Y_t^i \\ Y_t^i & \text{if } \hat{C}_t^i \leq Y_t^i < \tilde{C}_t^i \\ \hat{C}_t^i & \text{if } Y_t^i < \hat{C}_t^i. \end{cases} \tag{7.8}$$

As usual, we obtain the **realized total consumption expenditure** as $C_t = \sum_{i=1}^{N} C_t^i$ and calculate the realized growth factor c_t of aggregate consumption by

$$c_t = \frac{C_t}{C_{t-1}} - 1.$$

Finally, we calculate the realized distribution factors β_t^i of consumption for each household group under consideration and for the time t by

$$\beta_t^i = \frac{C_t^i}{C_t}.$$

Consideration of Inflation

We define inflation as an increase in projected consumer spending that exceeds economic growth. With given **inflation rates** r_t^{inf} we model the projected consumption of a household group by

$$\tilde{C}_t^i = \left(1 + r_t^{\text{inf}}\right)(1 + g_t)\, C_{t-1}^i.$$

The above algorithm is now applied to $\tilde{C}_t^i$.

Chapter 8
The Circular Flow Model with State

In this chapter, the state and its tax system are included in the modeling.

Proposition 4.7 showed that for $r \geq g \geq 0$ with $r > 0$ the Gini coefficients of wealth distributions in circular flow economies with constant parameters diverge toward ∞ for large times.

The inclusion of the state makes it possible to investigate whether and to what extent the inequality of wealth in an economy could be mitigated by tax regulations or subsidies.

8.1 Balance Equations of the Circular Flow Model with State

In the following, the state is included in the analysis as an economic actor. The state:

- Levies **taxes**
- Redistributes income and wealth by means of **tax revenues**, **government spending**, and **subsidies**
- Creates savings, preferably negative ones, which is then referred to as **public debt**

The circular flow model extended to include the state is shown in Fig. 8.1. Government employees are assigned to a state-owned company as part of the modeling process and thus included in the existing system.

As in Chap. 2, we begin the analysis of the circular flow model by deriving balance sheet equations.

J. Kremer, *Capital Income and Concentration of Wealth*, Springer Studies in Alternative Economics, https://doi.org/10.1007/978-3-658-50715-2_8

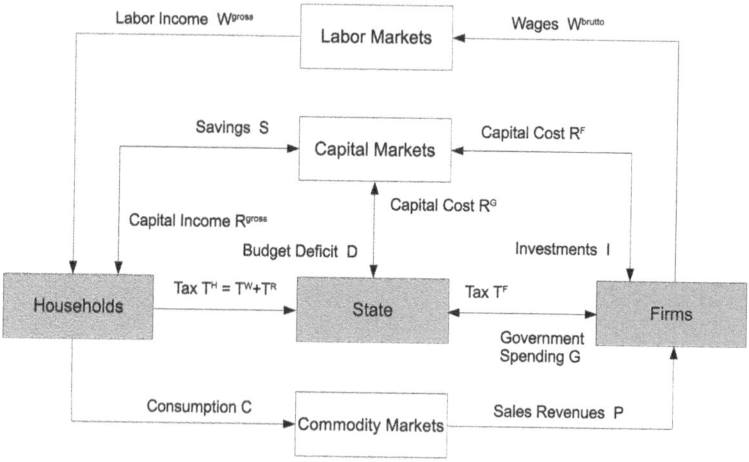

Fig. 8.1 Circular flow model for an economy with state

Households

Households pay **income tax** T^W on their **gross labor income** W^{gross}, so that they are left with their **net labor income**

$$W^{\text{net}} = W^{\text{gross}} - T^W. \tag{8.1}$$

They receive **gross capital income** R^{gross} based on their saved and invested wealth V, but they pay the **capital income tax** T^R on R^{gross}. The **net capital income** R^{net} is defined as the gross capital income R^{gross} reduced by the capital income tax T^R

$$R^{\text{net}} = R^{\text{gross}} - T^R. \tag{8.2}$$

We add up T^W and T^R to the **tax** T^H **of the households**

$$T^H = T^W + T^R. \tag{8.3}$$

Households use their **gross income** $W^{\text{gross}} + R^{\text{gross}}$ for **consumption expenditures** C, to pay taxes T^H and to invest **savings** S

$$W^{\text{gross}} + R^{\text{gross}} = C + T^H + S. \tag{8.4}$$

If components of T^H are negative, these are to be interpreted as **subsidies**. From (8.4), (8.1), (8.2), and (8.3), we obtain

$$W^{\text{net}} + R^{\text{net}} = S + C. \tag{8.5}$$

Firms

The income of companies is made up of **sales revenues** P, **government spending** G, and **investments** I.

Corporate expenditures consist of **gross wages** W^{gross}, the **corporate tax** T^F to be paid, and the share R^F of **capital costs** that is passed on to savers. The remaining share of the cost of capital becomes income for the commercial banks and is therefore part of the aggregated gross wages W^{gross}. Thus,

$$P + G + I = W^{\text{gross}} + R^F + T^F. \tag{8.6}$$

The gross wages paid by companies correspond to the gross labor income received by households. The term W^{gross} is therefore also used for gross wages.

State

The state levies taxes and grants subsidies. The taxes T^H are paid by households and the corporate tax T^F by firms. The sum

$$T^G = T^H + T^F \tag{8.7}$$

denotes the **government's aggregate tax revenue** in the context of our modeling. In addition, the government decides on **government spending** G. Capital costs must be paid for government debt. We use R^G to denote the share of the **government's capital cost** that is passed on to savers. The rest remains with the commercial banks and is therefore part of government spending G.

If the government's expenditure $G + R^G$ exceeds its revenue T^G, then a loan must be taken out for the difference D, the **budget deficit**. Thus,

$$D = G + R^G - T^G. \tag{8.8}$$

A positive value of D corresponds to the required credit volume. Negative values of D lead to a reduction in government debt or even to government savings.

Economic Relationships Between Households, Firms, and the State

Obviously,

$$P = C; \tag{8.9}$$

the sales revenues of companies correspond to the consumer spending of households.

Further,

$$R^{\text{gross}} = R^F + R^G, \tag{8.10}$$

The gross capital income R^{gross} of households is made up of the shares of the capital costs R^F of companies and the shares of the capital costs R^G of the state that are passed on to savers.

Then, with (8.3), (8.5), (8.6), (8.7), (8.8), (8.9), and (8.10), we obtain

$$\begin{aligned} I + D &= \left(W^{\text{gross}} + R^F + T^F - P - G\right) + \left(G + R^G - T^H - T^F\right) \\ &= \left(W^{\text{gross}} - T^H\right) + \left(R^F + R^G\right) - P \\ &= \left(W^{\text{gross}} - T^W\right) + \left(R^{\text{gross}} - T^R\right) - P \\ &= W^{\text{net}} + R^{\text{net}} - C \\ &= S, \end{aligned} \tag{8.11}$$

so that

$$S = I + D. \tag{8.12}$$

Aggregate household savings S are equal to the sum of aggregate corporate investment I and the government budget deficit D.

Let us denote the capital income R^{gross} of households by R^H

$$R^H = R^{\text{gross}}; \tag{8.13}$$

then (8.10) may be written as

$$R^H = R^F + R^G. \tag{8.14}$$

We define the **gross domestic product** Y analogous to (2.7) by

$$Y = C + S. \tag{8.15}$$

With (8.5) we now obtain the identities

$$\begin{aligned} Y &= W^{\text{net}} + R^{\text{net}} \\ &= C + S \\ &= P + I + D. \end{aligned} \tag{8.16}$$

The gross domestic product can therefore be characterized as the sum of all **net incomes** $W^{\text{net}} + R^{\text{net}}$ of households or alternatively as the sum $C + S$ of all savings amounts and consumption expenditure of households or finally as the sum $I + D + P$ of corporate investments, government budget deficit, and sales revenues of companies. **Economic growth** is defined, as before, as growth in gross domestic product.

Government Spending Ratio and Budget Deficit

The **government spending ratio** is defined as the share of government spending in gross domestic product

$$q = \frac{G + R^G}{Y} = \frac{D + T^G}{Y}, \tag{8.17}$$

where (8.8) was used. Thus,

$$D = qY - T^G. \tag{8.18}$$

For the circular flow economy algorithm, a government share must be specified, from which the budget deficit is then calculated using (8.18). Empirical values for Germany's government share in the past are between 40 and 50%.

Income Tax and Capital Gains Tax

The state levies taxes on both households and companies and invests them as government expenditure. In this way, taxes cause a redistribution of income and wealth.

Within the framework of the circular flow model, **income tax rates** τ_t^i are to be specified for each household group $i = 1, \ldots, N$ and for each year t. With the gross labor income $W_t^{\text{gross},i}$ of a group, the income tax $T_t^{W,i}$ for this group is calculated by

$$T_t^{W,i} = \tau_t^i W_t^{\text{gross},i}. \tag{8.19}$$

The net labor income $W_t^{\text{net},i}$ of the ith group is thus given by

$$\begin{aligned} W_t^{\text{net},i} &= W_t^{\text{gross},i} - T_t^{W,i} \\ &= \left(1 - \tau_t^i\right) W_t^{\text{gross},i}. \end{aligned} \tag{8.20}$$

The gross capital income $R_t^{\text{gross},i}$ for group i at a time t is the capital income from the wealth V_{t-1}^i at time $t-1$, i.e.,

$$R_t^{\text{gross},i} = r_t V_{t-1}^i, \tag{8.21}$$

where r_t denotes the capital return rate for the period $[t-1,\ t]$. The **capital gains tax**

$$T_t^{R,i} = \kappa_t^i R_t^{\text{gross},i} \tag{8.22}$$

is determined using tax rates κ_t^i for the capital gains. Thus, (8.22) specifies what proportion of a household group's capital income has to be paid in taxes. For the net capital income of a household group $R_t^{\text{net},i}$, the representation

$$\begin{aligned} R_t^{\text{net},i} &= R_t^{\text{gross},i} - T_t^{R,i} \\ &= \left(1 - \kappa_t^i\right) R_t^{\text{gross},i} \end{aligned} \tag{8.23}$$

is analogous to (8.20). The **total tax** $T_t^{H,i}$ of a household group i at time t is therefore

$$T_t^{H,i} = T_t^{W,i} + T_t^{R,i}. \tag{8.24}$$

Interest Transfer

We consider the net capital income and the net capital costs of a household group. The net capital income of the ith group is $R_t^{\text{net},i}$ at time t. If the aggregated net capital income R_t^{net} is allocated to the households as capital costs according to shares in total consumption, then the capital costs Π_t^i of the ith household group hidden in consumption are, analogous to Sect. 2.4, given by

$$\Pi_t^i = \frac{C_t^i}{C_t} R_t^{\text{net}} = \beta_t^i R_t^{\text{net}}.$$

As balance Z_t^i of capital income $R_t^{\text{net},i}$ and capital costs Π_t^i of the ith household group we obtain

$$Z_t^i = R_t^{\text{net},i} - \Pi_t^i. \tag{8.25}$$

We call Z_t^i the **interest transfer** of group i at time t. Here, too, $\sum_{i=1}^N Z_t^i = 0$, so that the gains of the groups with $Z_t^i > 0$ are financed by the groups with $Z_t^i < 0$.

8.2 Recursion Algorithm

For the recursion algorithm, the relationship (5.5)

$$V_t^i = V_{t-1}^i + S_t^i \tag{8.26}$$

is used in addition to the relationships shown in Sect. 8.1.

The amounts of wealth $\left(V_{-1}^1, \ldots, V_{-1}^N\right)$ of the N modeled household groups and government debt B_{-1} for time -1 have to be specified, as well as gross labor incomes $\left(W_0^{\text{gross},1}, \ldots, W_0^{\text{gross},N}\right)$ and consumption expenditures $\left(C_0^1, \ldots, C_0^N\right)$ for each group at time 0. Furthermore, a government spending ratio q, the rates of return on capital r_t, $t = 0, \ldots, T$, and the economic growth factors g_t, $t = 1, \ldots, T$, have to be given.

At the beginning of the simulation calculation, some initializations have to be made, such as the determination of the capital income $R_0^i = r_0 V_{-1}^i$ at time 0 or the calculation of the net labor income $\left(W_0^{\text{net},1}, \ldots, W_0^{\text{net},N}\right)$. The assumed income tax rates have an effect here. As part of the initializations, subsidies can also be incorporated into the algorithm as grants to earned income at time 0.

The following calculations are then carried out in a loop over the years from $t = 1$ to the final year $t = T$, assuming that all the required data, such as Y_{t-1}, V_{t-1}^i, etc., are available for the previous year $t-1$:

- The gross investment income $R_t^{\text{gross},i} = r_t V_{t-1}^i$ is used to calculate the capital gains taxes $T_t^{R,i}$ and thus the net capital income $R_t^{\text{net},i} = R_t^{\text{gross},i} - T_t^{R,i}$ of the households.
- With the factor g_t for economic growth, the current gross domestic product is $Y_t = (1 + g_t)\, Y_{t-1}$. This leads to the aggregate net labor income $W_t^{\text{net}} = Y_t - R_t^{\text{net}}$ of the economy.
- The aggregated net labor income is distributed to the household groups according to the rule chosen for the simulation, $W_t^{\text{net},i} = \alpha_t^i W_t^{\text{net}}$, where α_t^i denotes the associated distribution factors.
- Now the total net income can be calculated for each household group, $Y_t^i = W_t^{\text{net},i} + R_t^{\text{net},i}$.
- The next step is to determine the income tax $T_t^{W,i}$ that each household group has to pay. To do this, the gross earned income $W_t^{\text{gross},i}$ must be calculated from the corresponding net earned income $W_t^{\text{net},i}$ for each group. Now, the income tax is given by $T_t^{W,i} = W_t^{\text{gross},i} - W_t^{\text{net},i}$.
- The total tax to be paid by a household group is $T_t^{H,i} = T_t^{R,i} + T_t^{W,i}$.
- The rule selected for the simulation is now used to calculate the consumption expenditure C_t^i for each household group and then the distribution factors $\beta_t^i = C_t^i / C_t$.

- The interest transfer results in $Z_t^i = R_t^{\text{net},i} - \beta_t^i R_t^{\text{net}}$.
- The savings and the wealth of each group are given by $S_t^i = Y_t^i - C_t^i$ and $V_t^i = V_{t-1}^i + S_t^i$.
- The VAT to be paid by the companies for a VAT rate τ^F is given by $T_t^F = \tau^F C_t$.
- This determines the current tax revenue $T_t^G = T_t^H + T_t^F$ of the economy.
- With a specified government ratio q, the budget deficit $D_t = qY_t - T_t^G$ is calculated.
- The government debt is then given by $B_t = B_{t-1} + D_t$, and the cost of capital is $R_t^G = r_t B_{t-1}$.
- Government expenditure is therefore $G_t = D_t - R_t^G + T_t^G$.
- The companies' cost of capital is given by $R_t^F = R_t^{\text{gross}} - R_t^G$.
- Now, all the data required for year t are calculated, and the algorithm is iterated until the final year T of the simulation is reached.

8.3 Development of Wealth as Dynamical System

The results of this section are analogous to those of Sect. 5.2. In particular, we will see that the recursion algorithm of a circular flow economy with state can also be formulated as a dynamical system. To verify this, we use (8.16),

$$Y_t = C_t + S_t = W_t^{\text{net}} + R_t^{\text{net}}, \tag{8.27}$$

and (8.26),

$$V_t^i = V_{t-1}^i + S_t^i.$$

The following proposition extends Proposition 5.2 to include capital gains tax.

Proposition 8.1 *The wealth V_t^i of a household group i at time t is given by*

$$V_t^i = \sum_{j=1}^{N} a_t^{ij} V_{t-1}^j + b_t^i \tag{8.28}$$

with

$$a_t^{ij} = \left(1 + r_t\left(1 - \kappa_t^i\right)\right)\delta^{ij} - r_t\alpha_t^i\left(1 - \kappa_t^j\right) \tag{8.29}$$

$$b_t^i = \alpha_t^i Y_t - \beta_t^i C_t = \alpha_t^i S_t + \left(\alpha_t^i - \beta_t^i\right) C_t.$$

Here, δ^{ij} with $\delta^{ij} = 1$ for $i = j$ and $\delta^{ij} = 0$ for $i \neq j$ denotes the Kronecker symbol.

Proof First of all, using the capital gains tax rates κ_t^i, we have

$$\begin{aligned} R_t^{\text{net},i} &= R_t^{\text{gross},i} - T_t^{R,i} \\ &= r_t \left(1 - \kappa_t^i\right) V_{t-1}^i. \end{aligned}$$

With (8.26) and (8.27), we obtain

$$\begin{aligned} V_t^i &= V_{t-1}^i + S_t^i \\ &= V_{t-1}^i + W_t^{\text{net},i} + R_t^{\text{net},i} - C_t^i \\ &= V_{t-1}^i + r_t \left(1 - \kappa_t^i\right) V_{t-1}^i + W_t^{\text{net},i} - C_t^i. \end{aligned}$$

Furthermore,

$$\begin{aligned} W_t^{\text{net},i} &= \alpha_t^i W_t^{\text{net}} \\ &= \alpha_t^i \left(Y_t - R_t^{\text{net}}\right) \\ &= \alpha_t^i \left(Y_t - r_t \sum_{j=1}^{N} \left(1 - \kappa_t^j\right) V_{t-1}^j\right), \end{aligned}$$

and thus we find the recursion

$$V_t^i = \left(1 + r_t \left(1 - \kappa_t^i\right)\right) V_{t-1}^i - r_t \alpha_t^i \sum_{j=1}^{N} \left(1 - \kappa_t^j\right) V_{t-1}^j + \alpha_t^i Y_t - C_t^i,$$

as was to be shown. □

The wealth development of the household groups can thus be formulated as a dynamical system analogous to Corollary 5.3:

Corollary 8.2 *We have*

$$\mathbf{V}_t = \mathbf{A}_t \mathbf{V}_{t-1} + \mathbf{b}_t \tag{8.30}$$

with

$$\mathbf{V}_t = \begin{pmatrix} V_t^1 \\ \vdots \\ V_t^N \end{pmatrix}, \tag{8.31}$$

$$\mathbf{A}_t = \begin{pmatrix} a_t^{11} & \cdots & a_t^{1N} \\ \vdots & & \vdots \\ a_t^{N1} & \cdots & a_t^{NN} \end{pmatrix} \tag{8.32}$$

$$= I + r_t \left(I - \begin{pmatrix} \alpha_t^1 & \cdots & \alpha_t^1 \\ \vdots & & \vdots \\ \alpha_t^N & \cdots & \alpha_t^N \end{pmatrix} \right) \left(I - \begin{pmatrix} \kappa_t^1 & \cdots & 0 \\ \vdots & \ddots & \vdots \\ 0 & \cdots & \kappa_t^N \end{pmatrix} \right)$$

and

$$\mathbf{b}_t = \begin{pmatrix} b_t^1 \\ \vdots \\ b_t^N \end{pmatrix}. \tag{8.33}$$

Proof The assertions follow directly from Proposition 8.1. □

Assume that the capital return rates, the economic growth rates, and the capital gains tax rates are constant over time

$$r_t = r, \quad g_t = g, \quad \kappa_t^i = \kappa^i.$$

Furthermore, we assume time-constant allocation factors for the distribution of aggregate net labor income and for the distribution of total consumption among the household groups, i.e.,

$$\alpha_t^i = \alpha^i, \quad \beta_t^i = \beta^i.$$

Then, we have for all t

$$W_t^{\text{net},i} = \alpha^i \, W_t^{\text{net}}$$

and

$$C_t^i = \beta^i \, C_t.$$

The development of gross domestic product is given by

$$Y_t = (1+g)^t \, Y_0.$$

If we assume that total consumption C_t grows with the gross domestic product, then

$$C_t = (1+g)^t \, C_0.$$

We further have

$$R_t^{\text{net},i} = r\left(1-\kappa^i\right)V_{t-1}^i$$

for all t. Thus, the matrices $\mathbf{A}_t =: \mathbf{A}$ are constant over time, and the first line of (8.29) simplifies to

$$a^{ij} = \delta_{ij} + r\left(\delta_{ij} - \alpha^i\right)\left(1-\kappa^j\right).$$

Therefore,

$$\mathbf{A} = \begin{pmatrix} 1+r\left(1-\alpha^1\right)\left(1-\kappa^1\right) & \cdots & -r\alpha^1\left(1-\kappa^N\right) \\ -r\alpha^2\left(1-\kappa^1\right) & \cdots & -r\alpha^2\left(1-\kappa^N\right) \\ \vdots & \ddots & \vdots \\ -r\alpha^N\left(1-\kappa^1\right) & \cdots & 1+r\left(1-\alpha^N\right)\left(1-\kappa^N\right) \end{pmatrix}$$

$$= I + r\left(I - \begin{pmatrix} \alpha^1 & \cdots & \alpha^1 \\ \vdots & & \vdots \\ \alpha^N & \cdots & \alpha^N \end{pmatrix}\right)\left(I - \begin{pmatrix} \kappa^1 & \cdots & 0 \\ \vdots & \ddots & \vdots \\ 0 & \cdots & \kappa^N \end{pmatrix}\right),$$

where I denotes the unit matrix, and we have

$$b_t^i = \left(\alpha^i S_0 + \left(\alpha^i - \beta^i\right)C_0\right)(1+g)^t.$$

Analogous to Sect. 6.4 we obtain

$$\mathbf{V}_t = \mathbf{A}^{t+1}\mathbf{V}_{-1} + \sum_{j=0}^{t}\left(\mathbf{A}^j\mathbf{b}_0\right)(1+g)^{t-j}.$$

8.4 Rules for the Simulation Calculations

Distribution of Labor Income and Consumer Spending Among the Household Groups

The rules:

- For the allocation of net labor income to the household groups and
- For the development of consumption expenditure and its allocation to the household groups

is determined first. Thus, $x_{0.5}$ is the gross income, for which half of the top tax rate applies. In [20], this amount $x_{0.5}$ is called **semi-top tax rate income**. If $x_{0.5}$ is given, then k is determined by the condition

$$\frac{c_{\max}}{2} = \tau_{k,c}\left(x_{0.5}\right)$$

from which we conclude

$$\tfrac{1}{2} = e^{-\frac{k}{x_{0.5}}},$$

that is,

$$k = x_{0.5}\ln 2.$$

This leads to the following definition:

Definition 8.4 The **Beuth approximation** consists of using the Beuth function $\tau_{x_{0.5},c_{\max}}$ to define the tax rates. Thus, for a taxable income $x > 0$, the tax rates are modeled as

$$\begin{aligned}\tau_{x_{0.5},c_{\max}}(x) &= c_{\max}e^{-\frac{x_{0.5}\ln 2}{x}}\\ &= c_{\max}2^{-\frac{x_{0.5}}{x}},\end{aligned}$$

where $c_{\max}$ denotes the top tax rate and $x_{0.5}$ is the semi-top tax rate income.

By determining the parameters $x_{0.5}$ and $c_{\max}$, the Beuth functions can be used to approximate the tax functions for any years. The **tax function** defined using the Beuth approximation is given by

$$\tau_{x_{0.5},c_{\max}}(x)\,x,$$

the **net function**, which assigns the corresponding net amounts to the gross amounts $x > 0$, is

$$x - \tau_{x_{0.5},c_{\max}}(x)\,x = \left(1 - c_{\max}2^{-\frac{x_{0.5}}{x}}\right)x.$$

Figure 8.2 shows the graph of the tax rate function $\mathrm{sf}_{2021}(x)/x$ for the year 2021 together with the corresponding Beuth approximation. Figure 8.3 above shows the graphs of the tax function $\mathrm{sf}_{2021}(x)/x$ and the corresponding Beuth approximation $\tau_{x_{0.5},c_{\max}}(x)\,x$, and Fig. 8.3 below shows the corresponding net functions $x - \mathrm{sf}_{2021}(x)$ and $\left(1 - \tau_{x_{0.5},c_{\max}}(x)\right)x$.

For simulation calculations, the Beuth functions are used in the following representation:

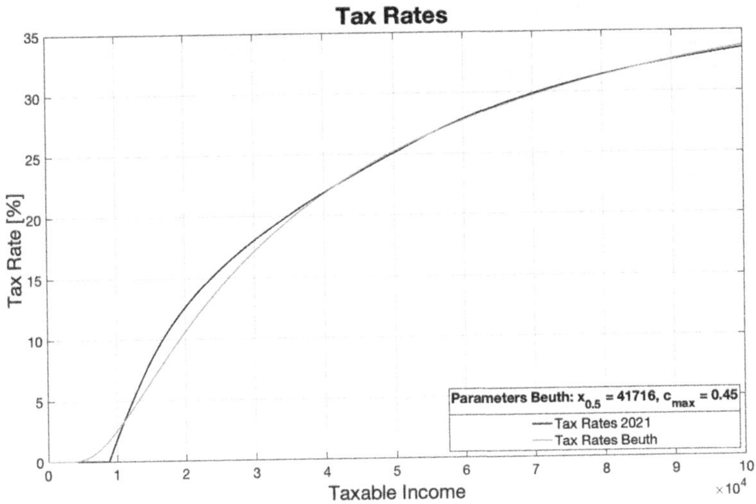

Fig. 8.2 Tax rates $sf_{2021}(x)/x$ and approximating Beuth function $\tau_{x_{0.5},c_{\max}}$

Definition 8.5 The **normalized Beuth functions** $z_{b,c} : (-\infty, 1] \to [0, c]$ are given by

$$z_{b,c}(x) = \begin{cases} 0 & \text{if } x \leq 0 \\ c2^{-b\left(\frac{1}{x}-1\right)} & \text{if } 0 < x \leq 1. \end{cases} \tag{8.34}$$

This assumes $b > 0$ and $c \in [0, 1]$.

For any c, $z_{b,c}(1) = c$, regardless of $b > 0$. For each of the N modeled household groups of an economy, let values $Q^1, \ldots, Q^N$ of a taxable variable Q be given, where $Q^0 = \max\left(Q^1, \ldots, Q^N\right) > 0$ is assumed. The variable Q can be the gross labor income or the gross capital income. The tax rate for Q^0 is denoted by c. Furthermore, let $c_{\max}$ be the top tax rate and $Q^{0.5}$ the value of Q at which the tax rate $c_{\max}/2$ occurs. Further, let b be defined by

$$b = \frac{Q^{0.5}}{Q^0}. \tag{8.35}$$

First, the tax rates are then determined using the Beuth approximation by

$$\tau_{Q^{0.5},c_{\max}}(x) = c_{\max}2^{-\frac{Q^{0.5}}{x}}.$$

On the other hand, using (8.35) we may write

$$\tau_{Q^{0.5},c_{\max}}(x) = \left(c_{\max}e^{-b}\right)2^{-b\left(\frac{Q^0}{x}-1\right)}.$$

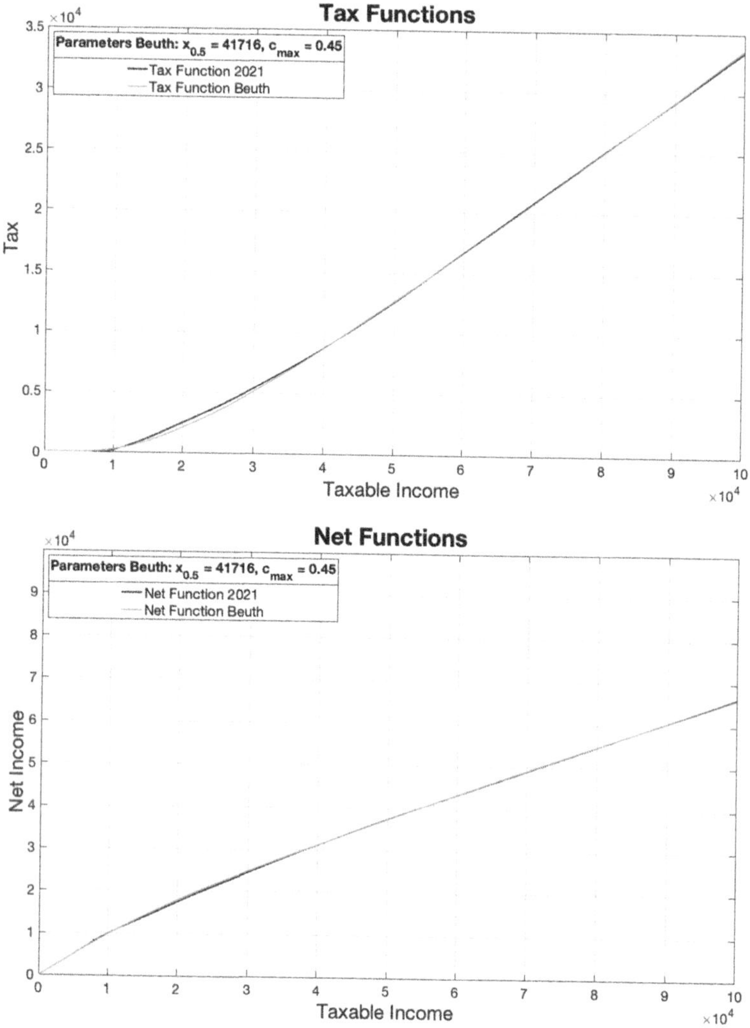

Fig. 8.3 Tax function and net function for the year 2021, each calculated with sf_{2021} and using the approximating Beuth function $\tau_{x_{0.5},c_{\max}}$

Thus, with

$$c = c_{\max} e^{-b},$$

we conclude

$$\tau_{Q^{0.5},c_{\max}}\left(Q^i\right) = ce^{-b\left(\frac{Q^0}{Q^i}-1\right)} = z_{b,c}\left(\frac{Q^i}{Q^0}\right).$$

Obviously,

$$z_{b,c}(1) = \tau_{Q^{0.5},c_{\max}}\left(Q^0\right) = c,$$

and

$$\begin{aligned} z_{b,c}\left(\frac{Q^{0.5}}{Q^0}\right) &= c2^{-b\left(\frac{Q^0}{Q^{0.5}}-1\right)} \\ &= c2^{-b\left(\frac{1}{b}-1\right)} \\ &= \tfrac{1}{2}c2^b \\ &= \tfrac{1}{2}c_{\max}. \end{aligned}$$

We are thus led to the following statement:

Lemma 8.6 *For each household group, let a value $Q^1, \ldots, Q^N$ of a taxable quantity Q be given with $Q^0 = \max\left(Q^1, \ldots, Q^N\right) > 0$. The tax rate for Q^0 is denoted by c and the top tax rate by $c_{\max}$. Let $Q^{0.5}$ be the value of Q for which the tax rate $c_{\max}/2$ occurs. With*

$$b = \frac{Q^{0.5}}{Q^0},$$

the tax rates for the Q^i are then defined by

$$z_{b,c}\left(\frac{Q^i}{Q^0}\right) = ce^{-b\left(\frac{Q^0}{Q^i}-1\right)},$$

where $z_{b,c}$ denotes the normalized Beuth function. □

The following example demonstrates the use of normalized Beuth functions to determine tax rates, income taxes, and net labor income.

Example 8.7 Let gross labor income

$$\left(W^{\text{gross},i}\right)_{i=1,\ldots,10} = (9,\ 15,\ 19,\ 23,\ 27,\ 30,\ 36,\ 42,\ 52,\ 96)$$

be given in the unit TE for an economy divided into ten household groups. With the tax function sf_{2021}, this results in the rounded net labor income

$$\left(W^{\text{net},i}\right)_{i=1,\ldots,10} = (9,\ 14,\ 17,\ 20,\ 23,\ 25,\ 29,\ 33,\ 39,\ 65) \tag{8.36}$$

in the unit TE.

The rounded tax rate calculated with the tax function sf_{2021} for an annual gross labor income of 96,000 E is 32%. Therefore, the parameter c of the normalized Beuth approximation is

$$c = 0{,}32.$$

The top tax rate of the tax function for 2021 is 45%, and this results in a rounded semi-top tax rate income of

$$W^{\text{gross},0.5} = 45.5\,\text{TE}.$$

Thus, we have

$$\frac{45\%}{2} = 22.5\% = \frac{sf_{2021}\left(1000 \cdot W^{\text{gross},0.5}\right)}{1000 \cdot W^{\text{gross},0.5}}.$$

With $W^{\text{gross},0} = \max\left(W^{\text{gross},1}, \ldots, W^{\text{gross},10}\right) = 96$, the parameter b of the normalized Beuth approximation is therefore estimated as

$$b = \frac{W^{\text{gross},0.5}}{W^{\text{gross},0}} = \frac{45.5}{96} = 0.47.$$

The tax rate τ^i for the ith gross labor income $W^{\text{gross},i}$ is now approximated by

$$\tau^i = z_{b,c}\left(\frac{W^{\text{gross},i}}{W^{\text{gross},0}}\right),$$

and the income tax to be paid is estimated as

$$T^{W,i} = \tau^i W^{\text{gross},i}.$$

The rounded net labor income calculated using the normalized Beuth approximation is thus finally calculated as

$$\begin{aligned}\left(W^{\text{net},i}\right)_{i=1,\ldots,10} &= \left(W^{\text{gross},i} - T^{W,i}\right)_{i=1,\ldots,10} \\ &= (9,\ 14,\ 17,\ 20,\ 23,\ 25,\ 29,\ 33,\ 39,\ 65),\end{aligned}$$

and this coincides with (8.36). △

Normalized Beuth functions are used in this book to model tax rates for the time-varying distributions of labor income. If a normalized Beuth function $z_{b,c}$ is selected, it is used to assign a tax rate of $c = z_{b,c}(1)$ to the highest labor income of each income distribution. The tax-free amount can be adjusted via the parameter b. The higher the value of b, the higher the resulting tax-free amount, see Fig. 8.4.

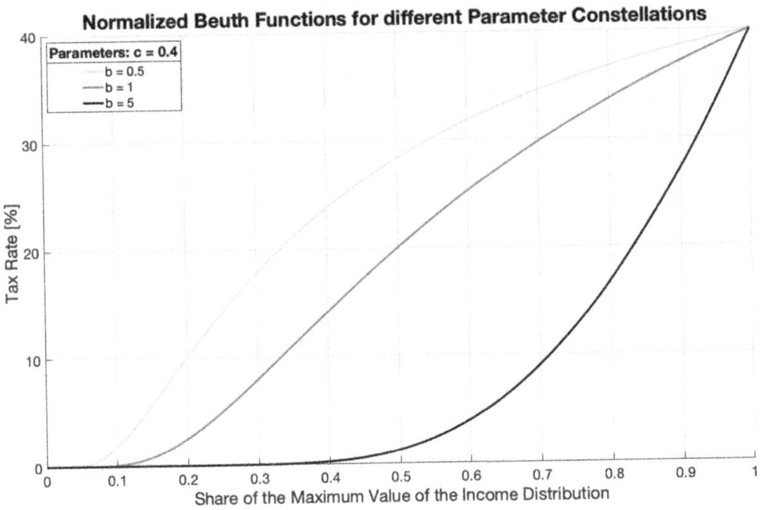

Fig. 8.4 Normalized Beuth functions for different parameter constellations

As part of the algorithm for a circular flow economy, it is necessary to determine the corresponding gross labor income from given net labor income. Let $W^{\text{net},1}, \ldots, W^{\text{net},N}$ be given net labor incomes with $W^{\text{net},0} = \max\left(W^{\text{net},1}, \ldots, W^{\text{net},N}\right) > 0$.

Then, because of $z_{b,c}(1) = c$, we have

$$W^{\text{net},0} = (1 - c)\, W^{\text{gross},0},$$

so that

$$W^{\text{gross},0} = \frac{W^{\text{net},0}}{1 - c}.$$

For realistic taxation of labor income, it can be assumed that the net function

$$f(x) = x - z_{b,c}\left(\frac{x}{W^{\text{gross},0}}\right) x,$$

which assigns the net income $f(x)$ to a taxable labor income $x > 0$, is strictly monotonically increasing. This means that the net amounts $W^{\text{net},1}, \ldots, W^{\text{net},N}$ corresponding to the gross amounts $W^{\text{gross},1}, \ldots, W^{\text{gross},N}$ are given as the uniquely determined zeros of the functions

$$x \mapsto W^{\text{net},i} - \left(1 - z_{b,c}\left(\frac{x}{W^{\text{gross},0}}\right)\right) x$$

for $i = 1, \ldots, N$.

Rules for the Taxation of Capital Income

According to (8.22) capital gains tax $T_t^{R,i}$ is given by

$$T_t^{R,i} = \kappa_t^i R_t^{\text{gross},i},$$

where κ_t^i denote the tax rates for capital gains. This section provides rules for determining these tax rates.

Rule: Capped Capital Gains Tax Rates

The capital gains tax in Germany at the time of publication of this book is determined as follows: If τ_t^i denotes the tax rate for the taxation of labor income of a household group i at time t, then the tax rates κ_t^i for the taxation of capital gains with $\tau_{\max} = 25\%$ are defined by

$$\kappa_t^i = \begin{cases} 0 & \text{if } R_t^{\text{gross},i} \leq 0 \\ \tau_t^i & \text{if } R_t^{\text{gross},i} > 0 \text{ and } 0 \leq \tau_t^i < \tau_{\max} \\ \tau_{\max} & \text{if } R_t^{\text{gross},i} > 0 \text{ and } \tau_t^i \geq \tau_{\max}. \end{cases} \tag{8.37}$$

If the tax rate τ_t^i for labor income is between 0 and $\tau_{\max}$, then this rate is also applied to capital income if it is positive. However, if τ_t^i is greater than $\tau_{\max}$, then the capital gains tax rate is set to $\tau_{\max}$ for positive capital income, i.e., capped by this value.

Rule: Determination of the Capital Gains Tax Using a Tax Function

For given parameters $b > 0$ and $0 \leq c \leq 1$, the tax rates κ_t^i for the investment income of any household group i at any point in time t are defined by

$$\kappa_t^i = z_{b,c}\left(\frac{R_t^{\text{gross},i}}{R_t^{\text{gross},0}}\right), \tag{8.38}$$

where $R_t^{\text{gross},0} = \max\left(R_t^{\text{gross},1}, \ldots, R_t^{\text{gross},N}\right) > 0$ is assumed. Figure 8.5 illustrates the normalized Beuth function $z_{b,c}$ with associated tax function $z_{b,c}(x)\,x$ and net function $\left(1 - z_{b,c}(x)\right)x$ for the case $b = 0.5$ and $c = 1$. In the case of $c = 1$, high capital income is taxed increasingly more heavily, up to full taxation of the highest capital income.

Because $R_t^{\text{gross},i} = r_t V_t^i$, (8.38) may for $r_t > 0$ also be written as

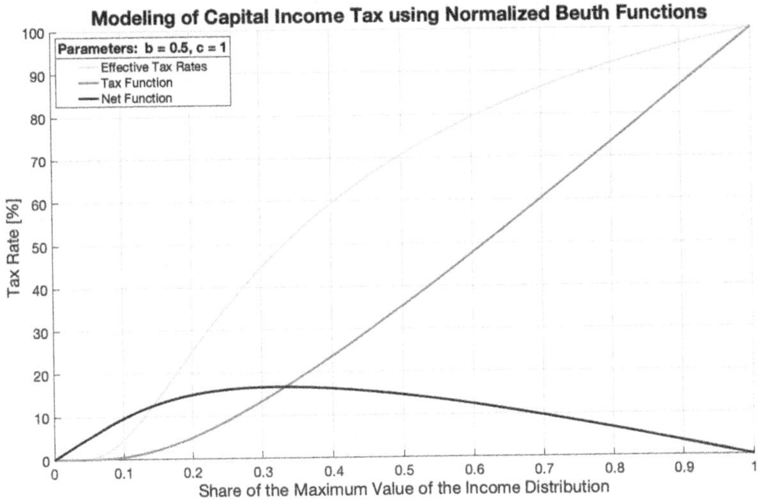

Fig. 8.5 Normalized Beuth function $z_{b,c}$ for modeling capital income tax rates with associated tax and net functions for the parameter constellation $b = 0.5$ and $c = 1$

$$\kappa_t^i = z_{b,c}\left(\frac{V_t^i}{V_t^0}\right), \tag{8.39}$$

where $V_t^0 = \max\left(V_t^1, \ldots, V_t^N\right) > 0$ is assumed.

In (8.38) and (8.39), c is a constant. Assume that the extent of inequality in the wealth distribution $V_t^1, \ldots, V_t^N$ of the economy has been reduced over time by appropriate taxation of capital income. The maximum tax rate c could then be lowered depending on the degree of inequality, and the Gini coefficient G_t of the wealth distribution could be used to assess the degree of inequality of wealth at a point in time t. For example, given a fixed $0 < \gamma \leq 1$, a maximum tax rate c_t could be defined by

$$c_t = \begin{cases} c & \text{falls } G_t > \gamma \\ \frac{1}{\gamma} G_t c & \text{falls } G_t \leq \gamma. \end{cases} \tag{8.40}$$

For $\gamma = 0.5$ in particular, this rule leads to

$$c_t = \begin{cases} c & \text{falls } G_t > 0.5 \\ 2G_t c & \text{falls } G_t \leq 0.5. \end{cases} \tag{8.41}$$

8.5 Example Simulations

This section presents the wealth developments of household groups in an economy for three example scenarios. For each scenario, the initial data specified in the appendix were used. All simulations were carried out using the parameter configuration $r = 5\%$ and $g = 2\%$. In addition, for all scenarios it was assumed that household groups which are no longer able to finance their projected consumption expenditures implement spending cuts down to the critical growth rate $\hat{g} = g/2$ before beginning to finance their consumption through borrowing. Details on the rules applied can be found in Chap. 7 and in Sect. 8.4.

The first simulation is based on the tax rules for labor income and capital income that were in effect in 2021. In the other simulations, capital income from large fortunes was entirely taxed away as long as the Gini coefficient of the wealth distribution exceeded 0.5. For Gini coefficients below 0.5, the maximum tax rate c_t was determined by Eq. (8.41) with $c = 1$.

In the final scenario, no taxes were levied on labor income. This third scenario is also compared to the first—that is, the simulation assuming the tax regulations valid in 2021. It turns out that government tax revenues in the third simulation are higher in the long run, despite the fact that no taxes are levied on labor income in this case.

For each simulation, the Lorenz curve and the Gini coefficient of the wealth distribution were calculated and plotted for the selected time points 0, 25 and 50. In addition, the simulated distributions of wealth, net labor income, and consumption expenditures at times $0, 10, \ldots, 50$ were rounded and presented in tabular form in units of TE.

1. Simulation Using 2021 Taxation Rules for Labor and Capital Income
Here, the taxation rules for labor income and capital income valid in the year 2021 are applied. Specifically, this means that the tax rate applicable to labor income is also used for taxing capital income, as long as it does not exceed 25%. If the labor income tax rate of a household group exceeds 25%, then capital income is taxed at a flat rate of 25%. Furthermore, in this first scenario, the upper groups are favoured in the distribution of total labour income among the household groups, see Sect. 7.1. The results of this simulation are shown in Fig. 8.6.

(a) The Gini coefficients of the wealth distributions increase over time and eventually exceed the value of one.
(b) After 50 years, half of all households are in debt.
(c) With the exception of the tenth household group, the net labor income of all other groups is eventually no longer sufficient to cover their consumption expenditures.
(d) Consumer spending must then be partially financed through capital income or debt.

Over time, the simulated economic development proves to be increasingly unbalanced.

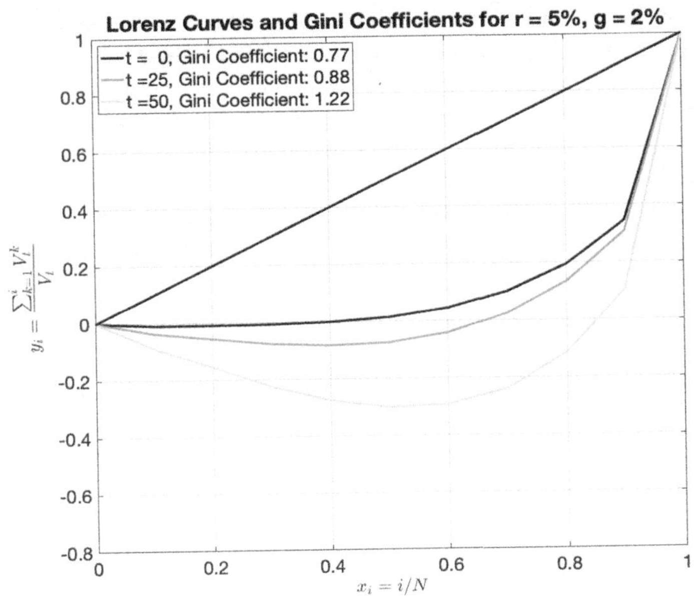

(a) Lorenz curves and Gini coefficients for wealth distributions

Jahre	HG_1	HG_2	HG_3	HG_4	HG_5	HG_6	HG_7	HG_8	HG_9	HG_10
0	-21	1	5	14	32	69	128	211	339	1479
10	-66	-9	-5	24	72	146	249	386	596	2431
20	-175	-73	-69	-10	72	209	389	621	969	3870
30	-385	-223	-226	-116	20	237	544	934	1506	6029
40	-760	-513	-536	-342	-119	214	707	1353	2280	9244
50	-1401	-1030	-1094	-769	-407	121	860	1925	3401	14006

(b) Wealth distribution

Year	HG_1	HG_2	HG_3	HG_4	HG_5	HG_6	HG_7	HG_8	HG_9	HG_10
0	9	14	17	20	23	25	29	33	39	65
10	7	12	15	18	23	26	31	37	45	80
20	6	10	13	16	21	26	33	41	52	97
30	4	8	11	15	20	25	34	45	60	118
40	3	6	9	13	18	25	35	49	70	144
50	2	5	7	11	17	24	36	53	80	176

(c) Net labor income

Year	HG_1	HG_2	HG_3	HG_4	HG_5	HG_6	HG_7	HG_8	HG_9	HG_10
0	10	13	16	17	19	20	23	26	30	43
10	11	15	18	20	23	24	28	32	37	52
20	12	16	20	22	27	30	34	39	45	64
30	13	18	22	25	30	36	42	47	54	78
40	15	20	25	27	33	40	51	57	66	95
50	16	22	27	30	36	44	62	70	81	116

(d) Consumer spending

Fig. 8.6 1. Simulation using 2021 taxation rules for labor and capital income

2. Simulation in Which Capital Income for High-Wealth Households Is Fully Taxed Away and the Dispersion of Labor Income Is Limited
Here, no cap on tax rates for capital income is imposed. Instead, capital income for households with high wealth is taxed at up to 100%. For determining the capital income tax, the rule (8.41) with $c = 100\%$ is applied. Furthermore, the dispersion of net labor income is limited. To achieve this, (7.1)—and consequently (7.2)—is assumed for the allocation of net labor income among the household groups, ensuring that each group's share of total net labor income remains constant over time. The results of this simulation are shown in Fig. 8.7.

(a) As a result of this rule, the Gini coefficient of the wealth distribution is reduced over time.
(b) With the exception of the first group, the wealth of all household groups increases over time.
(c) Net labor income is higher and the distribution of net labor income is more balanced compared to the first simulation.
(d) The distribution of consumer spending is more balanced than in the first example.

Over time, the Gini coefficient of the wealth distribution decreases. The wealthy household groups remain wealthy over time, but the wealth distribution is far more balanced than in the first simulation.

3. Simulation: Same as the Previous Scenario, But Without Taxes on Labor Income
In this scenario, the rules from the previous example apply, except that no taxes are levied on labor income. The results of this simulation are shown in Fig. 8.8.

(a) The Gini index changes very little in comparison with the previous simulation.
(b) In this scenario, the wealth of the household groups increase more sharply than in the previous one.
(c) The labor income of the upper half of the household groups is rising more sharply than in the previous simulation.
(d) The distribution of consumer spending is the same as in the previous example.

Here, in the 3rd simulation, an interesting aspect is the development of household tax payments compared to Scenario 1, which corresponds to the tax system in place in Germany in 2021. In that latter scenario, the evolution of household taxes is shown in Fig. 8.9. In contrast, if capital income from high wealth is fully taxed away and labor income is entirely exempt from taxation, the result is depicted in Fig. 8.10. In both figures, values are given in units of TE and rounded to a maximum of three decimal places.

We observe that government tax revenues are significantly higher over time when capital income from large fortunes is heavily taxed and labor income is not taxed at all, compared to the revenues generated under the current tax regulations.

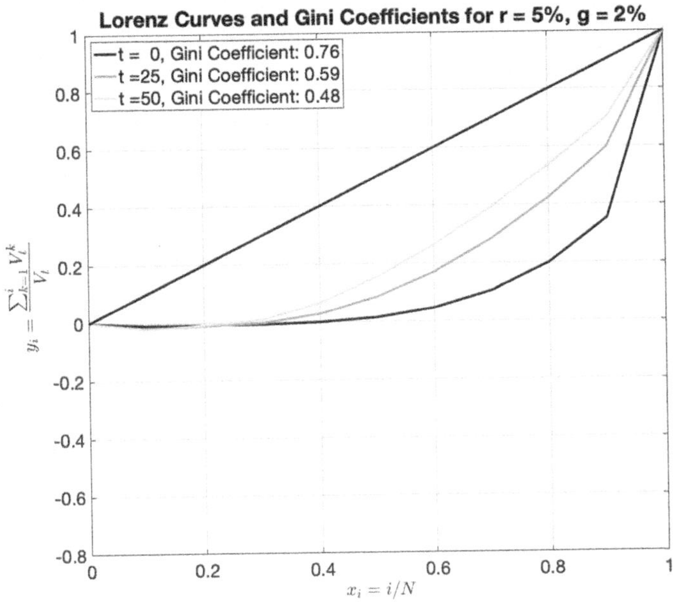

(a) Lorenz curves and Gini coefficients for wealth distributions

Year	HG_1	HG_2	HG_3	HG_4	HG_5	HG_6	HG_7	HG_8	HG_9	HG_10
0	-21	1	5	14	33	69	129	212	337	1426
10	-46	13	20	53	104	176	275	389	544	1655
20	-79	27	39	112	215	333	466	599	771	1906
30	-120	48	66	206	380	540	696	839	1022	2197
40	-168	80	109	351	600	794	965	1113	1305	2546
50	-219	132	178	557	876	1095	1276	1428	1631	2978

(b) Wealth distribution

Year	HG_1	HG_2	HG_3	HG_4	HG_5	HG_6	HG_7	HG_8	HG_9	HG_10
0	9	14	17	20	23	25	29	33	39	65
10	10	17	20	23	27	29	34	39	45	76
20	12	20	24	27	32	35	41	46	54	90
30	15	24	29	33	39	42	49	56	65	109
40	18	29	35	40	47	52	60	68	79	133
50	22	35	43	48	57	62	72	81	95	160

(c) Net labor income

Year	HG_1	HG_2	HG_3	HG_4	HG_5	HG_6	HG_7	HG_8	HG_9	HG_10
0	10	13	16	17	19	20	23	26	30	43
10	11	16	20	21	23	24	28	32	37	52
20	12	19	24	25	28	30	34	39	45	64
30	13	24	29	31	34	36	42	47	54	78
40	15	29	35	38	42	44	51	57	66	95
50	16	35	43	46	51	54	62	70	81	116

(d) Consumer spending

Fig. 8.7 2. Simulation in which capital income for high-wealth households is fully taxed away and the dispersion of labor income is limited

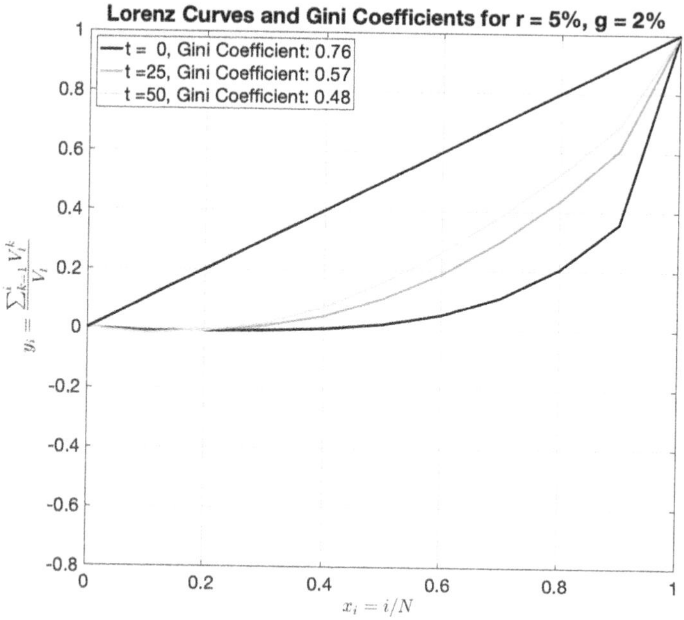

(a) Lorenz curves and Gini coefficients for wealth distributions

Year	HG_1	HG_2	HG_3	HG_4	HG_5	HG_6	HG_7	HG_8	HG_9	HG_10
0	-21	2	6	17	36	74	136	220	350	1457
10	-46	23	41	84	155	240	363	502	694	2003
20	-83	46	87	186	336	479	662	844	1089	2597
30	-135	77	153	343	596	795	1029	1248	1543	3270
40	-204	120	254	576	942	1191	1472	1724	2068	4053
50	-298	181	406	902	1379	1675	1999	2286	2686	4991

(b) Wealth distribution

Year	HG_1	HG_2	HG_3	HG_4	HG_5	HG_6	HG_7	HG_8	HG_9	HG_10
0	9	15	19	22	27	30	36	42	51	96
10	10	17	22	25	31	34	41	48	58	109
20	12	20	25	29	36	39	47	55	67	126
30	14	23	30	34	42	47	56	65	79	150
40	17	28	35	41	50	56	67	78	95	179
50	20	33	42	48	59	66	79	92	112	211

(c) Net labor income

Year	HG_1	HG_2	HG_3	HG_4	HG_5	HG_6	HG_7	HG_8	HG_9	HG_10
0	10	13	16	17	19	20	23	26	30	43
10	11	16	20	21	23	24	28	32	37	52
20	12	19	24	25	28	30	34	39	45	64
30	13	24	29	31	34	36	42	47	54	78
40	15	29	35	38	42	44	51	57	66	95
50	16	35	43	46	51	54	62	70	81	116

(d) Consumer spending

Fig. 8.8 3. Simulation: same as the previous scenario, but without taxes on labor income

Year	HG_1	HG_2	HG_3	HG_4	HG_5	HG_6	HG_7	HG_8	HG_9	HG_10	Sum
0	0.124	0.826	1.637	2.412	3.946	5.165	7.781	10.904	16.06	48.27	97.124
10	0.015	0.236	0.654	1.28	2.732	4.349	7.736	12.329	20.172	66.405	115.91
20	0.001	0.039	0.185	0.482	1.523	3.259	7.271	13.643	25.272	91.879	143.55
30	0	0.002	0.03	0.138	0.623	2.075	6.387	14.769	31.622	127.8	183.44
40	0	0	0.002	0.024	0.201	1.075	5.193	15.676	39.615	178.61	240.4
50	0	0	0	0.002	0.053	0.435	3.91	16.469	49.931	250.71	321.51

Fig. 8.9 Simulation 1: Development of total taxes $T^H = T^R + T^W$ of the household groups under the taxation rules applicable in 2021

Year	HG_1	HG_2	HG_3	HG_4	HG_5	HG_6	HG_7	HG_8	HG_9	HG_10	Sum
0	0	0	0	0	0	0.002	0.162	1.316	5.227	70.2	76.907
10	0	0	0	0.001	0.108	0.856	3.564	8.478	17.37	97.314	127.69
20	0	0	0	0.104	1.586	5.017	11.709	20.075	32.983	126.72	198.19
30	0	0	0.008	0.899	6.229	13.343	23.906	35.123	51.54	159.89	290.94
40	0	0	0.079	3.603	14.951	25.773	39.768	53.465	73.274	198.43	409.34
50	0	0.001	0.432	9.217	27.021	40.81	57.407	72.992	95.685	236.8	540.36

Fig. 8.10 Simulation 3: Development of total taxes $T^H = T^R + T^W = T^R$ of the household groups, assuming that capital income from high wealth is taxed at a high rate and that no taxes are levied on labor income, $T^W = 0$

8.6 Concluding Remarks

In the circular flow models presented in this book, capital income proves to be a powerful driver of wealth redistribution—from the broad majority of less affluent households toward a minority of wealthier ones. To quantify the redistribution mechanism, the concept of interest transfer was introduced or used in Sects. 2.4, 3.5, and 8.1. This transfer is always active, but in particular, under the conditions $r \geq g \geq 0$ and $r > 0$ in a circular economy with constant parameters, it has the unpleasant consequence of leading to wealth concentration within the economy.

When the state and its tax system are included in the analysis—as is done in Chap. 8—it becomes evident that heavy taxation of capital income from large fortunes can halt or even reduce this concentration.

In contrast, Debraj Ray argues in his article [18] that there is no causal relationship between $r > g$ and increasing inequality:

> You need something else to get at rising inequality. What then, explains the marked and disconcerting rise of inequality in the world today? Capital, in the physical and financial sense that Piketty uses it, has something to do with it. But it has something to do with it because it is a vehicle for accumulation. It is probably the principal vehicle for accumulation by the top 1% or the top 0.01%, simply because there are generally limits on how high the compensation to human capital can be in any generation. It is hard enough to make a few hundred thousand dollars in annual labor income, and reaching the million-dollar mark (let alone tens of millions) is far harder and riskier. But physical capital—land and financial assets—can be steadily and boundlessly accumulated. In this sense Piketty is right in turning the laser on capital. But, as I said, it's just a vehicle.

Appendix: Initial Data for the Simulation Calculations of the Book

Data for the Simulations in Chaps. 3 and 4

The simulation studies presented in Chaps. 3 and 4 examine how model economies evolve over time in the absence of government and tax systems, under different constellations of capital return rates and economic growth. The following initial data are used for the simulations in the specified chapters:

Wealth

The initial data for the wealth of the household groups at time 0 are specified as follows:

$$\left(V_0^1, \ldots, V_0^{10}\right) = (10,\ 17,\ 46,\ 61,\ 83,\ 114,\ 172,\ 245,\ 451,\ 1220)\,.$$

Labor Income

The following values are used as initial data for labor income:

$$\left(W_0^1, \ldots, W_0^{10}\right) = (10,\ 18,\ 22,\ 26,\ 29,\ 33,\ 38,\ 44,\ 53,\ 76)\,.$$

J. Kremer, *Capital Income and Concentration of Wealth*, Springer Studies in Alternative Economics, https://doi.org/10.1007/978-3-658-50715-2

Consumer Spending

The following values are assumed as initial data for consumer spending:

$$\left(C_0^1, \ldots, C_0^{10}\right) = (10,\ 18,\ 22,\ 26,\ 27,\ 29,\ 33,\ 37,\ 40,\ 45)\,.$$

All data given are annual values in the unit TE.

Concentration of Wealth

From the given initial data we compute

$$\begin{aligned}\left(\alpha^1, \ldots, \alpha^{10}\right) &= (0.0287,\ 0.0516,\ 0.0630,\ 0.0745,\ 0.0831,\ 0.0946,\ 0.1089,\\ &\qquad 0.1261,\ 0.1519,\ 0.2178)\,,\\ \left(\beta^1, \ldots, \beta^{10}\right) &= (0.0348,\ 0.0627,\ 0.0767,\ 0.0906,\ 0.0941,\ 0.1010,\ 0.1150,\\ &\qquad 0.1289,\ 0.1394,\ 0.1568)\,,\\ \left(\upsilon^1, \ldots, \upsilon^{10}\right) &= (0.0041,\ 0.0070,\ 0.0190,\ 0.0252,\ 0.0343,\ 0.0471,\ 0.0711,\\ &\qquad 0.1013,\ 0.1864,\ 0.5043)\,.\end{aligned}$$

This implies

$$\begin{aligned}&\left(\alpha^1 - \beta^1, \ldots, \alpha^{10} - \beta^{10}\right)\\ &= (-0.0062, -0.0111, -0.0136, -0.0161, -0.0110, -0.0065, -0.0061,\\ &\quad -0.0028,\ 0.0125,\ 0.0610)\end{aligned}$$

and

$$\begin{aligned}&\left(\upsilon^1 - \alpha^1, \ldots, \upsilon^{10} - \alpha^{10}\right)\\ &= (-0.0245, -0.0445, -0.0440, -0.0493, -0.0488, -0.0474, -0.0378,\\ &\quad -0.0248,\ 0.0346,\ 0.2866)\,.\end{aligned}$$

Together with (3.32), it follows that both Δ_1^i and Δ_2^i are negative for $i = 1, \ldots, 8$, while they each have positive values for $i = 9,\ 10$. According to Proposition 3.14, this implies that under the conditions $r \geq g \geq 0$ and $r > 0$, a concentration of wealth occurs in the model economy in the sense of Definition 3.12.

For $g = 5\%$ and $r = 2\%$, however, we find

$$\left(\frac{1}{g-r}\Delta_1^1 + \frac{\alpha^1}{g}S_0, \ldots, \frac{1}{g-r}\Delta_1^{10} + \frac{\alpha^{10}}{g}S_0\right)$$

$$= (3,\ 6,\ 7,\ 8,\ 78,\ 149,\ 185,\ 256,\ 466,\ 1094)\,.$$

Thus, the expressions

$$\frac{1}{g-r}\Delta_1^k + \frac{\alpha^k}{g}S_0$$

are positive for all $k = 1, \ldots, 10$, and it follows from Proposition 3.19 that under this parameter constellation no concentration of wealth is observed in the sense of Definition 3.12.

Data for the Simulations in Chap. 8

The simulation calculations presented in Chap. 8 allow us to examine how model economies evolve over time under different constellations of capital return and economic growth, based on initial data and various rules for the taxation of labor and capital income.

The initial data required for the model calculations include the wealth, gross labor income, and consumption expenditures of each household group.

Wealth

Data from the German Institute for Economic Research (Deutsches Institut für Wirtschaftsforschung, DIW) are used as the basis for the wealth distribution. In a discrete-time context, the initial data for wealth are interpreted as values for time -1 and are assumed as follows for the model calculations presented here:

$$\left(V_{-1}^1, \ldots, V_{-1}^{10}\right) = (-19,\ 0,\ 3,\ 11,\ 27,\ 61,\ 117,\ 196,\ 318,\ 1404)\,.$$

Gross Labor Income

The starting point is the following net labor incomes provided by the DIW:

$$\left(W_0^{\text{net},1}, \ldots, W_0^{\text{net},10}\right) = (9,\ 14,\ 17,\ 20,\ 23,\ 25,\ 29,\ 33,\ 39,\ 65)\,.$$

The normalized Beuth approximation (8.34) with the parameters $b = 0,47$, $c = 0,32$ was used to calculate the gross labor income from the above net labor income, which leads to the following results:

$$\left(W_0^{\text{gross},1}, \ldots, W_0^{\text{gross},10}\right) = (9,\ 15,\ 19,\ 22,\ 27,\ 30,\ 36,\ 42,\ 51,\ 96)\,.$$

Consumer Spending

In Chap. 8, the following values are used as initial data for consumer spending:

$$\left(C_0^1, \ldots, C_0^{10}\right) = (10,\ 13,\ 16,\ 17,\ 19,\ 20,\ 23,\ 26,\ 30,\ 43)\,.$$

All values provided are expressed in units of TE.

Concentration of Wealth

Within the framework of the discretized and rule-based circular flow economies presented in Chap. 8, it is not possible to calculate limits for the evolution of economic variables—particularly the development of wealth among household groups. However, for different parameter constellations, it is possible to identify trends, as demonstrated by the simulations in Sect. 8.5. These simulations, for example, show that the Gini coefficient of wealth distribution could be reduced over time through high taxation of capital income from large fortunes.

Bibliography

1. Abraham R, Marsden JE, Ratiu T (2002) Manifolds, tensor analysis and applications, 3rd edn. Springer, Berlin
2. Acemoglu D (2008) Introduction to modern economic growth. Princeton University Press, Princeton
3. Aghion P, Howitt P (1997) Endogenous growth theory. MIT Press, Cambridge
4. Aghion P, Howitt P (2009) The economics of growth. MIT Press, Cambridge
5. Barro RJ, Sala-i-Martin X (2004) Economic growth. Prentice-Hall of India, Delhi
6. Grabka MM (2021) Einkommensungleichheit stagniert langfristig, sinkt aber während der Corona-Pandemie leicht. Wochenbericht 18, DIW
7. Grabka MM, Halbmeier C (2019) Vermögensungleichheit in Deutschland bleibt trotz deutlich steigender Nettovermögen anhaltend hoch. Wochenbericht 40, DIW
8. Grabka MM, Westermeier C (2014) Anhaltend hohe Vermögensungleichheit in Deutschland. Wochenbericht 9, DIW
9. Grabka MM, Westermeier C (2015) Reale Nettovermögen der Privathaushalte in Deutschland sind von 2003 bis 2013 geschrumpft. Wochenbericht 34, DIW
10. Keen S (2022) The new economics. Polity Press, Cambridge
11. Krause U, Nesemann T (2012) Differenzengleichungen und diskrete dynamische Systeme. 2. Auflage, De Gruyter
12. Kremer J (2008) Dynamische Analyse – Die Untersuchung des langfristigen Verhaltens von Ökonomien, in Luderer B (Hrsg) Die Kunst des Modellierens. Springer, Berlin
13. Kremer J (2013) Grundlagen der Ökonomie. 2. Auflage, Metropolis
14. Kremer J (2016) Geld ohne Schuld – Geldsysteme und Vollgeldreform. Metropolis
15. Mankiw NG (2015) Yes, $r > g$. So What?. Am Econ Rev Papers Proc 105(5):43–47
16. Piketty T (2015) The economics of inequality. Harvard University Press, Cambridge
17. Piketty T (2017) Capital in the twenty-first century. Harvard University Press, Cambridge
18. Ray D (2015) Nit-Piketty: a comment on Thomas Piketty's capital in the twenty first century. CESifo Forum 16(1):19–25
19. Schröder C et al (2020) MillionärInnen unter dem Mikroskop: Datenlücke bei sehr hohen Vermögen geschlossen – Konzentration höher als bisher ausgewiesen. Wochenbericht 29, DIW
20. Schwarz DE (2017) Herleitung einer Einkommensteuerfunktion durch Modellierung mit Differentialgleichungen, Berichte aus der Mathematik, Physik und Chemie 2/2017, Beuth-Hochschule für Technik Berlin

J. Kremer, *Capital Income and Concentration of Wealth*, Springer Studies in Alternative Economics, https://doi.org/10.1007/978-3-658-50715-2

The manufacturer's authorised representative in the EU is Springer Nature Customer Service Centre GmbH, Europaplatz 3, 69115 Heidelberg, Germany. If you have any concerns regarding our products, please contact ProductSafety@springernature.com

Printed and bound by CPI Group (UK) Ltd, Croydon, CR0 4YY
11/10/2026
02231452-0001